ENERGY SCIENCE, ENGINEERING AND TECHNOLOGY

GAS BIOFUELS FROM WASTE BIOMASS

PRINCIPLES AND ADVANCES

ENERGY SCIENCE, ENGINEERING AND TECHNOLOGY

Additional books in this series can be found on Nova's website under the Series tab.

Additional e-books in this series can be found on Nova's website under the e-book tab.

ENERGY SCIENCE, ENGINEERING AND TECHNOLOGY

GAS BIOFUELS FROM WASTE BIOMASS

PRINCIPLES AND ADVANCES

ZHIDAN LIU
EDITOR

nova publishers

New York

Library of Congress Cataloging-in-Publication Data

Gas biofuels from waste biomass : principles and advances / editors, Zhidan Liu, Laboratory of Environment-Enhancing Energy (E2E), College of Water Resources and Civil Engineering, China Agricultural University, Beijing, China.
    pages cm. -- (Energy science, engineering and technology)
  Includes bibliographical references and index.
  ISBN 978-1-63483-192-5 (hardcover)
  1. Biomass gasification. 2. Hydrolysis. 3. Fermentation. I. Liu, Zhidan.
  TP339.G37 2014
  665.7'76--dc23
                        2015021922

Published by Nova Science Publishers, Inc. † *New York*

CONTENTS

PREFACE

Waste biomass includes agricultural residues, livestock wastes, municipal wastes and industrial organic wastes. It should be utilized; otherwise, it will cause the pollution of water, soil and even the atmosphere. Gas biofuels have attracted growing attention as a renewable and clean energy carrier. Gas biofuels include biogas, biohydrogen and its mixture i.e., biohythane, which can be produced via anaerobic fermentation or other processes from waste biomass. This book focuses on the principles of gas biofuels in terms of types of biofuels, biomass species, reactor configuration and production pathway. A number of books focus on the production of biogas or biohydrogen alone. In comparison, this book emphasizes the interactions and common knowledge of both. In addition, the potential of new technologies, such as microbial electrochemical technologies, and two-stage fermentation on gas biofuel production are highlighted and specifically discussed based on the authors' research basis. This book provides a state-of-the-art technological insight into the production of gas biofuels from waste biomass.

Specifically, this book consists of three parts. In Part I, the principles for gas biofuels production from waste biomass, including biogas production (Chapter 1) and biohydrogen production (Chapter 2). Part II focuses on the technical advances on gas biofuels production. Pre-treatment of biomass was firstly introduced in Chapter 3, whereas the advances of biogas production from high-solid wastes were discussed in Chapter 4 and Chapter 5. In comparison, biohydrogen production is reviewed not only through dark fermentation (Chapter 6) but also emerging microbial electrochemical technology (Chapter 7). The co-production of biohydrogen and biomethane is reviewed in Chapter 8. In addition to the utilization of carbon and hydrogen stored in biomass, nutrients recycling through algae technology is discussed in Chapter 9. Part III discusses the scale-up and industrialization of biofuels. An industrial case is introduced to analyse the bottlenecks and perspectives for development of gas biofuels.

This work was finally supported by Natural Science Foundation of China (21106080), the National Key Technology Support Program of China (2014BAD02B03), the Chinese Universities Scientific Fund (2012RC030), and NSFC-JST Cooperative Research Project (21161140328). This book will provide a comprehensive text on the science of production of gas biofuels (biogas, biohydrogen and its mixture) from waste biomass. The book is intended for academia (professors, students, etc.), industry (biofuel companies), and government agencies. This book should serve as a reference resource for university and industry scientists in the area of gas biofuels and waste management.

PART I. PRINCIPLES FOR GAS BIOFUELS PRODUCTION FROM WASTE BIOMASS

In: Gas Biofuels from Waste Biomass   ISBN: 978-1-63483-192-5
Editor: Zhidan Liu   © 2015 Nova Science Publishers, Inc.

Chapter 1

BASIC CONCEPT 1:
BIOGAS PRODUCTION VIA ANAEROBIC DIGESTION

Qianqian Sun, Yuxin Wang and Shuang Liu*
China Agricultural University, China

ABSTRACT

Biogas fementation relies on anaerobic digestion as the main technological procedure. It is a system of engineering integrated wastes treatment, biogas production and resources utilization to achieve ecological recyceing on evaluation index, affecting factors. The factors including temperature,pH, additives, depressors, as well as the three stages of biogas fementation which are hydrolytic acidification phase, hydrogen-producing acetogenic phase and methane-producing phase.The process, engineering apparatus,biogas transmission and distribution are mentioned concisely.

Keywords: biogas fermentation, acetogenic bacteria, methane bacteria, anaerobic condition

1. INTRODUCTION

As the energy shortage and the environmental pollution become more serious, seeking alternative clean energy has become an important energy developing strategy around the world. As an important part of renewable energy in China, large and medium sized biogas project has experienced significant development over the past few years. Biogas projects rely on anaerobic digestion as the main technological procedure, which is a system engineering integrated wastes treatment, biogas production and resources utilization to achieve ecological

---

* Corresponding Author address: Department of Agricultural Environment & Energy Engineering, College of Water Resources and Civil Engineering, China Agricultural University, Beijing,China, Email:meller@163.com

recycling and agricultural sustainable development. . In the following paragraphs, the principle, condition and process of biogas production via anaerobic digestion are introduced.

2. PRINCIPLE OF BIOGAS FERMENTATION

Biogas fermentation means the process of decomposing and converting various sorts of complex organic compounds in the fermented tank into methane and carbon dioxide and other substances by the effects of anaerobic microorganisms (including microorganisms with oxygen) in the absence of oxygen molecules, which also known as anaerobic digestion. Under normal circumstances, methane(CH_4) accounts for about 60%, and carbon dioxide (CO_2) accounts for about 40%, in addition, there is a small amount of hydrogen (H_2), nitrogen gas (N_2), carbon monoxide (CO_2), hydrogen sulfide (H_2S) and ammonia (NH_3) etc., [3].

Biogas fermentation is a complex, multi-steps reacting process under the combined action of different microorganisms, which is mainly divided into four steps: hydrolysis (liquefaction), acidification, acetoxylation and methanation. And there are three major groups of bacteria involved in the fermentation which are hydrolytic acidogenic bacteria, hydrogen-producing acetogenic bacteria and methanogenic bacteria bacteria.

First of all, hydrolytic bacteria and fermentative bacteria secrete a specific cell exoenzyme, catalyzing the hydrolysis of complex macromolecular organic compounds to smaller ones [17]. And then, fermentative bacteria will use hydrolyzates generated in the previous step to further ferment into specific fermentation products, such as acetic acid, hydrogen, carbon dioxide, and so on, and these fermentation products can be directly used by methanogenic bacteria to form methane and carbon dioxide.

While other fermentation products such as some alcohols and highly volatile fatty acids, need to be further used by methanogenic bacteria and converted into biogas, and usually this is achieved by symbiotic bacteria of methanogenic bacteria to decompose.

The process of biogas fermentation is divided into three successive stages, which are hydrolytic acidification phase, hydrogen-producing acetogenic phase and methane-producing phase. The main process of biogas fermentation as shown in Figure 1.

In the first stage of the hydrolysis and acidification phase, hydrolytic exoenzyme is generated by fermentative bacteria namely acidogenic bacteria which are mostly facultative anaerobic heterotrophic bacteria and specific anaerobic bacteria [21]. Under the action of these bacteria, complex macromolecules, insoluble organic compounds such as carbohydrates, fats, proteins, etc. are hydrolyzed to small dissolved organic molecules by the catalyzed oxidation of exoenzyme, and then transferred into microbial cells to further decompose generating volatile fatty acids, alcohols, aldehydes.

In the early period of hydrolysis acidification phase, due to the rate of carbohydrate decomposing into organic acids faster than that of protein decomposing to ammonia, accumulation of acids will bring about pH value of fermented liquid decrease rapidly; sometimes pH value can reach below 5. In the later period of hydrolysis acidification, with the decreasing of carbohydrate, decomposing of organic acids and breaking down of nitrogenous organic compound, alkaline substances such as ammonia, amines, and carbonate begin to accumulating and pH value of fermented liquid is gradually increased to 6.6-6.8. For the reason of strong adaptability of acid producing bacteria, a few minutes to several hours it can breed a generation.

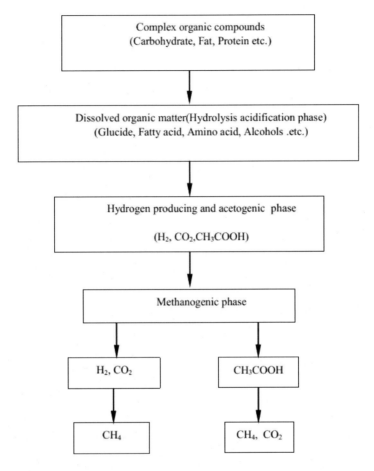

Figure 1. The main process of biogas fermentation [26].

In the second stage of hydrogen-producing acetogenic phase, acetogenic bacteria, such as glue acetic acid bacteria, some clostridium, etc., can decompose higher fatty acids to generate acetic acid and hydrogen. Furthermore, long-chain fatty acids can also be generated by fermentation bacteria in the decomposition of fatty acids, such as stearic acid. In the decomposition of proteins, aromatic acids, such as indole acitic acid generated fermentation bacteria. These acids are also decomposed by hydrogen-producing acetogenic bacteria in the second stage and generate acetic acid and hydrogen.

Under the action of anaerobic digestion bacteria, most organic acids and alcohols produced in the decomposition of complex organic compounds cannot be directly used by methane producing bacteria except formic acid, acetic acid and methyl alcohol, and they must be decomposed by acetogenic bacteria to convert into acetic acid, hydrogen and carbon dioxide [25]. Following are examples of the main reaction processes:

Propionic acid

$$CH_3CH_2COOH + 2H_2O \rightarrow CH_3COOH + 3H_2 \tag{1}$$

Butyric acid

$$CH_3CH_2CH_2COOH + 2H_2O \rightarrow 2CH_3COOH + 2H_2 \tag{2}$$

Lactic acid

$$CH_3CHOHCOOH + H_2O \rightarrow CH_3COOH + CO_2 + 2H_2 \tag{3}$$

In third stage of methane-producing phase, the formation of methane is actualized by a group of archaebacteria including hydrogen-eating methanogenic bacteria and acetic acid-eating methanogenic bacteria are physiologically specialized in converting acetic acid (acetate), CO_2 and H_2, etc. to methane. This process is implemented by two types of function entirely different methane producing bacteria, one is responsible for generating CH_4 from acetic acid or acetate decarboxylation, and the other one uses H_2 to deoxidize CO_2 back to methane. The former accounts for about 2/3 of the total, and the latter accounts for about 1/3. Their reaction formulas are as follows respectively.

$$4H_2 + CO_2 \rightarrow CH_4 + 2H_2O \tag{4}$$

$$4HCOOH \rightarrow CH_4 + 3CO_2 + 2H_2O \tag{5}$$

$$CH_3COOH \rightarrow CH_4 + CO_2 \tag{6}$$

$$CH_3COONH_4 + H_2O \rightarrow CH_4 + NH_4HCO_3 \tag{7}$$

Methanogenic bacteria exist widely in the sediment and other extreme anaerobic environment. For example, mthanosarcina exposes to air will soon die; its number of half-life is about 4 minutes. By determination, methanogenic bacteria cannot survive under the environment where oxidation reduction potential is more than -0.33V. Methane bacteria is very strict to pH value, generally suitable range is 6.8-7.8 and the optimum range within 6.2-7.2. Due to poor adaptability to temperature, methane bacteria, 1-2 degrees increasing of the temperature may make the destruction of digestion process. Methane bacteria propagation speed is very slow, usually takes 4-6 days to breeding on generation to breeding generation [6]. Normal process of anaerobic digestion should keep the forming velocity of acid-producing and methane-producing at equilibrium.

Although theoretically the anaerobic digested process can be divided into three stages above, these three stages are carried out simultaneously, in the biogas digesters with the absence of oxygen, and maintaining a certain degree of dynamic balance. Once the dynamic balance has been destroyed by factors such as temperature, pH, organic load, carbon nitrogen ratio of fermentation feedstock, oxidation-reduction potential, etc., the methane-producing stage will be inhibited firstly, as a result, it will lead to accumulation of the lower fatty acids and abnormal changes in anaerobic processes, and even lead to stagnation of the whole anaerobic digestion process.

3. FERMENTATION RAW MATERIAL AND EVALUATION INDEX

3.1. Fermentation Raw Material

Any biomass materials which are mainly composed of carbohydrates, proteins, fats, cellulose and hemicellulose can be used as the object of anaerobic fermentation. According to the differences in content of carbon and nitrogen, raw materials of biogas fermentation can be divided into two general categories: One is the raw material containing more carbon nutrition, such as straw etc, and the other one is raw materials containing more nitrogen nutrition, such as animal excreta etc. Different materials have differences in degradation characteristics because of their different components, and thus is the difference of gas generation rate. Table 1 is experienced gas production data sheet of conventional materials [20].

Table 1. Gas production data sheet of conventional materials

Material kind	The concentration of dry matter TS/%	theoretical gas production /(m^3/kg)
Caw dung	17-20	0.25-0.35
Pig manure	20-25	0.27-0.45
Fowl manure	10-29	0.3-0.8
Cereal straw	86	0.2-0.5
Sewage treatment sludge	--	0.2-0.7
Paper-making waste water	--	0.2-0.3

3.2. Evaluation Indexes of Raw Materials for Biogas Fermentation

The nature of the fermentation raw materials depend on the physicochemical properties of themselves, which decides the differences in the anaerobic treatment process, fermentation period, and biogas production rate [13]. In order to be able to accurately determine the content of organics in the fermentation materials and gas production potential, we usually adopt the following indicators to evaluate the fermentation raw materials.

1.   Total Solid (TS), also known as dry matter. Making a certain amount of fermentation material dried to constant weight in the oven in the temperature range of 103-105℃, and total solid is what we got. Total solid includes soluble solid and insoluble solid. The content of total solid in the fermentation raw materials is expressed as a percentage commonly, which is calculated as follows:

The content of total solids in fermentation raw materials

$$TS(\%) = \frac{W_2}{W_1} \times 100\%$$

(8)

Wherein,

W_1—Weight of the sample before drying (g);

W_2—Weight of the sample after drying, which is the weight of the total solid (g).

Total solids in a liquid sample also known as the total solids concentration. Total solids in a liquid sample can also use the g / L or mg / L to say, its formula is as follows:

$$TS(\%) = \frac{W}{V} \times 1000$$

(9)

Wherein,

W —Weight of the sample after drying, which is the weight of the total solid (mg);

V—Volume of the liquid sample (L).

2. Suspended Solid (SS) refers to solids in the liquid samples that could not pass through the filter. It can be obtained from the total solids and dissolved solids difference or from the measurement directly, which is often expressed as g/L or mg/L.

3. Volatile Solid (VS), Volatile solids refer to solids that can be volatilized at high temperature of 550°C, residues that cannot be volatilized are called ash. Generally volatilized solids are regarded as organics, but CO_2 and NH_3 will be decomposed and released when burning carbonate and ammonium salts, so in fact the VS has a small amount of inorganic compounds.

4. Biochemical Oxygen Demand (BOD) refers to oxygen consumption that microorganisms degrading organic matters in the water and reaching the stabilization required under aerobic conditions. The more organic matters in water samples, the greater the BOD. We take 20°C, 5-day BOD as measurement of organic matters can be decomposed by microorganisms in water samples normally, expressed by BOD_5, and its unit is mg/L.

5. Chemical Oxygen Demand (COD) refers to oxygen consumption that organic matter and strong oxidizing agent potassium dichromate react in water under a certain condition. When potassium dichromate is used as an oxidant agent, almost all organic matter is oxidized in water samples, and in this case the amount of oxygen consumption is known as chemical oxygen demand, expressed by COD.

4. Affecting Factors of Fermentation

Microorganisms of biogas fermentation require for suitable living conditions, where temperature, pH, oxidation reduction potential and other factors should all meet certain requirements.

4.1. Temperature

According to temperature difference, biogas fermentation can be divided into three types: normal temperature fermentation, mesophilic fermentation and thermophilic fermentation.

It is generally acknowledged that temperature for thermophilic fermentation is between 50-65°C, for mesophilic fermentation is 30-45°C, and for normal temperature fermentation is 10-25°C. Normal temperature fermentation refers to the process of biogas anaerobic fermentation under normal air temperature or water temperature [22].

The optimum temperature for mesophilic fermentation is 35°C -38°C. If the temperature is lower than 32°C or higher than 40°C, the efficiency of anaerobic digestion would reduce distinctly. The optimum temperature for thermophilic fermentation is 50°C-55°C. What's more, thermophilic fermentation produces approximately twice as much methane as mesophilic fermentation does.

However, although more biogas can be produced through thermophilic fermentation in a shorter period of time compared with mesophilic fermentation, it takes more energy to keep the digestive apparatus operating at a higher temperature [8]. Thus, on the net energy yield perspective, thermophilic fermentation is not economical.

Thanks to thermophilic digestion being especially sensitive to temperature change, a sudden rise or fall of temperature can significantly affect biogas yield. Thus, when designing a biogas digester, certain measures must be taken to ensure the digestive apparatus working at a constant temperature with the range of temperature change floating under 3°C per hour.

4.2. pH Value

Microorganism is active within a certain pH range. In the process of anaerobic digestion, the rise and fall of pH is not only influenced by external factors, but also by the increase or decrease of some products formed during the metabolic process of organics. The accumulation of organic acid would lead to the drop of pH, while, the increase of ammonia caused by the degradation of nitrogenous organic compound would lead to the increase of pH.

4.3. Carbon Nitrogen Ratio (C/N) of Fermenting Materials

Since biogas fermenting process is the process of cultivating anaerobic microorganisms, the culture medium should be considered in respect of carbon, nitrogen, phosphor, microelements and other nutrient substances that are essential for the growth of microorganisms.

Carbon nitrogen ratio (C/N) of fermenting materials refers to the ratio of organic carbon and nitrogen in raw materials. Since the content of carbon and nitrogen are quite different in different organic substances, before putting them into the digester, the carbon nitrogen ratio should be at a suitable level.

Generally speaking, 20-30:1 (C/N) is considered as an advisable carbon nitrogen ratio for fermentation. When C/N is 35:1, the yield would decrease dramatically. According to the percentage of carbon and nitrogen in each raw material, the carbon nitrogen ratio of mixed material can be calculated by the following formula:

$$C/N = \frac{C_1 W_1 + C_2 W_2 + \cdots + C_i W_i + \cdots + C_n W_n}{N_1 W_1 + N_2 W_2 + \cdots + N_i W_i + \cdots + N_n W_n}$$

(10)

Wherein,

i —indicate number i material, $i = 1 \cdots n$;

C_i —the percentage of carbon in material i, %;

N_i — the percentage of nitrogen in material i, %;

W_i — the weight of material i, kg.

4.4. Additives

The anaerobic fermentation process can be accelerated by many substances. The substance that can accelerate the destruction of organic matters and consequently raise biogas yield is called fermentation additive. To raise the yield of biogas and ensure the fermentation processed smoothly, various additives, including some enzymes, inorganic salts, organics and rare earth, are discovered through researches.

4.5. Depressors

Substances that can depress the activities of fermenting microorganisms are given an umbrella name as depressors. Normally, there are hardly toxic substances in crop straws, but a fair quantity of medicine would enter into fecal sewage during the disinfection or epidemic prevention process in livestock and poultry farms, besides, organic sewage from factories is also a source of toxic substances [24].

Fermentation would be depressed if the concentration of volatile acid in the biogas digester is too high. Also, if the concentration of ammonia nitrogen is too high, biogas fermentative bacteria would be depressed or even killed. Pesticides, especially the toxic ones, with strong bactericidal effects, can destroy the normal biogas fermentation process with even preciously little dose. Allowable concentrations for some salts and heavy metal compound during biogas fermentation are as shown in Table 2.

Table 2. Allowable concentration for some salts and heavy metal compound during biogas fermentation

Serial number	Compound	Allowable concentration(mg/L)
1	CuS	700
2	$Ni(NO_3)_2 \cdot 6H_2O$	200
3	$NaNO_3$	100
4	$NiSO_4 \cdot 7H_2O$	300
5	$NiCl_2 \cdot 6H_2O$	500
6	$CuCl_2 \cdot H_2O$	700
7	$Cr(OH)_3$	1000
8	NaCl	30000
9	$CuCl_2 \cdot H_2O$	700

5. TYPES OF BIOGAS FERMENTATION

A complete biogas fermentation project, regardless of its size, includes the following steps: collection of raw materials (sewage), pretreatment, biogas digester, post-processing of output, purification and storage of biogas.

Biogas fermentation can be divided into different types based on fermentation temperature, feeding method, concentration of fermentation liquid and different stages of fermentation.

5.1. Classification According to the Temperature

According to the temperature, there are three types of fermentation treatment including thermophilic fermentation, mesophilic fermentation and normal temperature fermentation.

1. In thermophilic fermentation, the range of fermenting temperature is 50-65°C and the degradation of organics is fast. It is suitable for treatment of higher concentration organic sewage.
2. In mesophilic fermentation, the range of fermentation temperature is 30-45°C, where the biogas yield is stable with higher methane conversion efficiency.
3. In normal temperature fermentation, the fermentation temperature varies with the environment temperature (10-25°C), but its biogas yield is instable and the methane conversion efficiency is low.

5.2. Classification According to Feeding Method

According to feeding method, there are about four types of fermentation treatment including semi continuous fermentation, batch fermentation, continuous fermentation and two-step fermentation.

1. Under semi continuous fermentation, a batch of raw materials is put into the biogas digester for one cycle, and then refeed the digester with another batch of raw materials.
2. Under batch fermentation, biogas fermentation is processed in a normal way, and when the biogas yield begins to fall, small amount of materials are added. Namely, feeding and outputting are implemented periodically. This method produces biogas in a balanced way and has strong applicability.
3. Under continuous fermentation, certain amount of materials is added continuously or with very short intervals after the biogas fermentation begins to operate normally. Being able to produce biogas in a balanced way and of higher conversion efficiency, this method is commonly used to dispose organic sewage.
4. Under two-step fermentation, the acid production stage and the methane production stage of biogas fermentation are separated into two apparatuses. Biogas yield and the degradation efficiency of volatile solid could be greatly improved through two-step fermentation. Besides, the solid waste decomposition period is shortened and the disposal of waste can be done in a more effective way.

6. PROCESSES OF BIOGAS FERMENTATION

A complete biogas fermentation project, regardless of its size, all includes the following steps: collection of raw materials (sewage), pretreatment, biogas digester, post-processing of output, purification and storage of biogas.

1. Pretreatment of Raw Materials

In biogas fermentation, in order to increase the biogas yield and shorten the start-up time, raw materials are always pretreated. Generally, the stage before putting materials into an anaerobic digester is called the pretreatment stage. To prevent the potential problems of fermenting devices caused by fibrous material agglomeration, raw materials are usually need to be smashed, mixed and heated before pumped into the digester.

2. Anaerobic Digestion

Biogas fermentation is completed with the cooperation of various microorganisms, and the biochemical degradation process of each organics is extremely complex. Microorganisms of biogas fermentation require for suitable living conditions, where temperature, pH, oxidation reduction potential and other factors should all meet certain requirements [2].

Anaerobic reactor (digester) is the core of biogas engineering. According to the structure type, biogas anaerobic reactors are mainly divided into Upflow Anaerobic Sludge Bed (UASB) anaerobic reactor, Internal Circulation (IC), anaerobic reactor, Constantly Stirred Tank Reactor (CSTR), Plug Flow Reactor(PFR), Upflow Solids Reactor (USR) and so on.

Taking Upflow Solids Reactor (USR) as an example, it is a kind of anaerobic reactor which has the advantages of simple structure and especially suitable for high suspended solid material.

After the high-density materials to be processed entering the anaerobic reactor from the bottom of the pipe, it will contact with the anaerobic activated sludge in the reactor and the high concentration of organic compounds in fermentation liquid will be rapidly digested and decomposed [10]. There is no need of setting the three-phase separator in the USR reactor, no need of inverse sludge system or any stirring devices. The undecomposed organic matters and anaerobic activated sludge particles will strand in the reactor by gravity. The supernatant will be discharged from the upper anaerobic reactor. This can get higher Solid Retention Time (SRT) than Hydraulic Retention Time (HRT) and Microbial Retention Time (MRT), so as to achieve higher solid organic matter decomposition rate. Schema of USR is as shown in Figure 2.

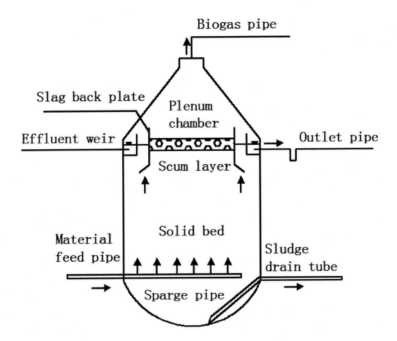

Figure 2. Schema of USR (Upflow Solids Reactor).

Anaerobic reactor should meet the need of anti-permeability and gas tightness during the process of anaerobic digestion. In case of the accidents of ultrahigh pressure or negative pressure, a voltage stabilizer should be set. It is also important to pay attention to the influence of temperature on the reactor. Heating and insulation measures should be taken accordingly to make sure that the fermentation materials maintain a favorable temperature and the reactor works normally, especially in cold winter [12]. In addition, effective anti-corrosion measures should be taken to ensure the safe operation of the reactor and lengthen its equipment life.

7. The Purification and Transportation of Biogas

Normally, at a stable and common fermentation phase, the biogas is composed of 50%-70% of methane, 30%-40% of carbon dioxide and small amounts of gases including carbon monoxide, water vapor, hydrogen, hydrogen sulfide, oxygen and nitrogen [9]. The biogas produced after fermentation cannot be used in industry and daily life directly before the purification.

7.1. Dewatering

The output biogas from the digester usually carries lots of water vapor, which is especially common in mesophilic fermentation at the present stage. During the transportation of the biogas, the vapors would condensate due to the change of temperature and pressure, which would increase the flow resistance in the pipes. At the same time, heat value of the biogas will decrease and corrosion of the pipes will happen. There are mainly two types dewatering methods.

1. Physical condensation. Condensation is the easiest physical method to remove the vapor in the methane. Condensate water, usually 1°C below dew point temperature, is removed in the heat exchange system by a cooler. To achieve a better effect of condensation, the methane is usually firstly compressed before being released to the needed pressure. Generally speaking, the lower the dew point temperature is, the higher the pressure needs to be. There are also many different kinds of condensations, including the demister, the cyclone separator and the moisture capturer.
2. Chemical exsiccation is usually achieved under relatively high temperature, with the appliance of desiccant like silica gel, alumina, magnesia, and water-absorptive salts etc... Two sets of apparatus are commonly used when applying desiccant, one for absorption and desiccation, and other for regenerative action. The granules of water-absorptive salts are usually filled in the adsorption tower so that when mixed methane goes through the apparatus, the vapor in the methane can be absorbed.

7.2. Desulfuration

A small amount of hydrogen sulfide is created during anaerobic fermentation, because hydrogen sulfide can react with most metals, compressors, gas storage tanks and engines would suffer from erosion. Therefore, in the purification of biogas, it is usually firstly desulfurized. At the present stage, desulfurization is usually carried out through two methods, biological oxidation and chemical oxidation.

1. Biological oxidation mainly refers to the method of converting hydrogen sulfide into elemental sulfur through the metabolism of microbial sulfur bacteria, including photosynthetic bacteria, denitrifying bacteria and colorless sulfur bacteria. The metabolic product can be released out of the cell by sulfur bacteria during the

metabolism process, so if oxygen concentration meets the standard for the limiting factors of sublimation reaction, the purpose of desulfurization can be successfully achieved through metabolism.

2. Chemical oxidation mainly refers to the chemical reactions between peroxides and sulfides. The basic ways of chemical desulfurization are shown in the following:

   (a) Digester in situ desulfurization

It has the advantages of low investment cost, low energy consumption easy operate and maintain and hydrogen sulfide is avoided to enter biogas pipelines. Disadvantages of digester in situ desulfurization are higher hydrogen sulfide concentration (100-150ppmv) of the output and higher operation cost (ferric salt).

   (b) Bed of Fe_2O_3/$Fe(OH)_3$

It has the advantages of higher hydrogen sulfide removal rate (over 99%) and low investment cost. But it has higher operation cost, large amount of regenerative heat, reaction surface decreasing with regenerations and releasing toxic dust.

   (c) Pressurized washing

It has the advantages of lower hydrogen sulfide concentration (15ppmv) in treated biogas and low cost if the water doesn't need regeneration, at the same time, CO_2 in the biogas can be removed. But high pressure results in higher operation cost and absorption tower is easy to be jammed.

   (d) Membrane separation

It has the advantages of higher hydrogen sulfide removal rate (over 98%) and CO_2 can be removed at the same time. But it need complex apparatus and result in higher operation and maintenance cost.

   (e) Biological desulphurization

It has the advantages of higher hydrogen sulfide removal rate (over 97%) and lower investment cost. But it will cause introduction of O_2 and N_2 possibly in biogas storage tank.

7.3. Decarbonation

The content of carbon dioxide in the methane after anaerobic digestion amounts to 30%-40%. When using methane as fuel (especially engine fuel), the heat of combustion of the methane would be affected by carbon dioxide, which would further affect the output power and other performances of the engine [16].

1. Condensation Adsorption

The main ingredients of the mixed gas are methane and carbon dioxide. Since the freezing point of carbon dioxide is higher than methane, when cooling the mixed gas, the carbon dioxide in the mixed gas would be firstly separated out. This method has a higher adsorption rate for carbon dioxide as well as a good purification effect for methane (over 97%); however, the disadvantages are high investment cost, high energy consumption, complex operation and strict requirements.

2. Chemical Adsorption

Amino solvents and heated potassium nitrate solution can both react with carbon dioxide. In addition, amino solvents like monoethanol amine, diethanol amine and triethanol amine are very active and can react with general acid gases [1]. Thus, in the purification process of methane, amino solvents not only absorb carbon dioxide, but also absorb a certain amount of hydrogen sulfide. Now, MDEA is widely used in actual production, because it can reduce the burden of regeneration and erosion, and improve the anti-degradation ability. The advantages of this method include relatively low cost of adsorbents, economy and easy to operate.

8. Basic Device of Biogas Fermentation

8.1. Gas Storage Device

Biogas storage system comprises a gas storage tank and flow meter etc. Gas storage tank is generally arranged near the anaerobic reactor. Considering safety, it must set some safety devices, such as fire arrestor and prevent excessive Inflating and pumping.

Figure 3. Dual-mode flexible dry gas cabinet.

Generally, the low pressure gas storage tank and a high-pressure gas storage tank are both adopted. Low pressure gas storage tank refers to the working pressure below 10 kPa. For example, low pressure dry gas holder can be divided into the rigid structure of gasholder and flexible air bag type gas holder. A flexible air bag dry type gas storage tank is also called double membrane dry gas holder (Figure 3), which is made of outer film, inner film, reinforced concrete foundation, air pump, pressure regulating device, volume table (range finder), methane gas leak detectors and other components [25]. The space between the inner and outer membrane is used as the pressure chamber regulating air storage. Compared with the wet gas holder, dry gas holder without water sealing, is not affected by the climate, convenient installation, stable operation, small amount of maintenance, low operation cost and long service life.

8.2. Integrated Biogas Fermentation Device

With the continuous development of new materials, new workmanship and new technology, the large and medium scale biogas engineering at home and abroad adopts the idea of the integrating the fermentation tank and the gas holder with the instillation of double-film structure at the upper part of the reactor. This technology has the advantage of fast construction and small occupation area, which can reduce the engineering investment [14].

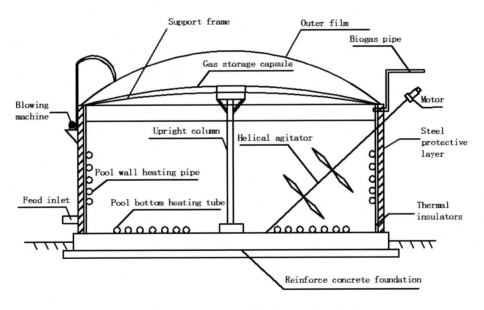

Figure 4. Gas production and storage integrated anaerobic fermentation device structure.

The tank of integrated biogas fermentation device (Figure 4) can use ordinary steel tank, Lipp tank, enamel plate assembly tank or reinforced concrete structure [7]. It is usually covered with double film structure on the top. According to the film function and characteristics, the storage film is divided into inner film and outer film. The space between the outer film and inner film is pumped to ensure constant air pressure and the air layer has a better heat insulation function. Biogas is stored in the space between inner film and fermentation tank liquid. According to the current workmanship level, the volume of integrated biogas fermentation device can reach 3000m^3.

9. COMPREHENSIVE UTILIZATION OF BIOGAS RESIDUE AND SLURRY

In spite of biogas, anaerobic digestion also produces a large quantity of residue and slurry. With the direction in utilization being farm-oriented, domestic and overseas researches mainly focus on study of their nutrient contents and utilization.

Biogas slurry has metabolites of anaerobic fermentation bacteria, high content of organic matter, plenty of elements like available N, P, K, among which the physiological activators and nutrient contents are fertilizers with high nutrient potential for farmland. In addition, a variety of water-soluble nutrients are contained in biogas slurry [5]. After diluting it 10 times, stirring and stewing, the supernatant liquor can be used as a spray fertilizer, which can increase the chlorophyll content of plant, promote crop photosynthesis and growth, and boost crop yields. After being diluted 10 times or so and stirred, the biogas slurry can be directly applied to the crop roots as root fertilizer and will play an important role in crop growth.

In anaerobic fermentation process, various active substances are produced by microorganisms in the biogas slurry. With the combined effects of those substances and some nutrients and the microelements on sprouting and growing, the biogas slurry can promote the germination of seeds and growth of seedlings. Therefore, biogas slurry is frequently used to improve the germination rate, the disease and insect resistance properties of the seeds.

In agricultural production, biogas residue mainly serves as nutrient medium to provide nutrient substance for crop growth. The composition of biogas residue is similar with that of biogas slurry; it also contains lots of essential nutrient substances which plants need, such as nitrogen, phosphorus and potassium, etc. Biogas residue can be used as base fertilizer that widely used in the cultivation of plants. While in the plant growth phase, biogas residue also can be used as topdressing to improve the quality of vegetables [11]. The application of biogas residue can increase the content of soil organic matter and improve the soil quality.

Under the global economic circumstances, scientific and technological level has become the basic element of regionals and countries' core competitiveness. Biogas project is an important way to realize the resource utilization of agricultural wastes and reduce environmental pollution. With the rapid development of large-size livestock farming and the request of continuous improvement of the environmental protection, the poultry farm waste treatment must be solved [13]. The ecological agricultural model driven by large-and-medium-scale biogas project is of good benefits in environment, society and economy.

ACKNOWLEDGMENTS

This work was Natural Science Foundation of China (21106080), the Chinese Universities Scientific Fund (2012RC030), the National Key Technology Support Program of China (2014BAD02B03), and NSFC-JST Cooperative Research Project (21161140328).

REFERENCES

[1]   Tang Xuemeng, Chen Li, Dong Renjie. Survey analysis and advice on large and medium Size. Biogas Plants in Beijing. *Journal of Agricultural Mechanization Research*, 2012, 34(3):206-211.

[2]   Wang Yuxin, Su Xing, Tang Yanfen. Status analysis and countermeasures of large and medium scale biogas plants in Beijing rural areas. *Transactions of the Chinese Society of Agricultural Engineering*, 2008,24(10):291—295.

[3]   Tu Yunzhang, Wu Zhaoliu. The Development report of large and medium scale biogas industrial. *Solar Energy*, 2012 (2):3-25.

[4]   McCarty P L., Smith DP. *Environ. Sci. Technol.*, , 20(12):1200-1206.

[5]   Speece,R.E, translated by Li Yaxin. Anaerobic biotechnology for industrial wastewaters. *China Building Industry Press*, 2014,4.

[6]   Chen Yongsheng, Zhu Dewen, Qu Haoli. Materials pretreatment process and equipment technology of the large and medium sized biogas projects. *Chinese Agriculture Mechanization*, 2010(06): 73-78.

[7]   Liu Xiangli, Yan Xuechang. The Application of Lipp technology on sewage treatment project. *GuangZhou Chemical Industry and Technology*, 2011,39(13):129-130.

[8]   Zhang Yijian. The intelligent control of biogas to produce gas-and-liquid integration anaerobic reaction jar. *Agricultural Technology and Equipment*, 2010(22):30-32.

[9]   Hu Mingcheng, Long Tengrui, Li Xuejun. New technical developments in the biogas desulfurization technology. *China Biogas*, 2005, 23(1):17-20.

[10] Tian Qiangdong, Qiang Jian. The fourth technology lectures on large and medium scale biogas project: the pretreatment of biogas project and the transmission and distribution system. *Renewable Energy*, 2003,(2):53-56.

[11] Song Chengfang, Shan Shengdao and Zhang Miaoxian. Study on concentration of biogas slurry from livestock and poultry wastes using membrane technology. *China Water and Wastewater*, 2011,27(3):84-86.

[12] He Rongyu. The strategy and issue on the running of large and middle-scale biogas projects in winter of Beijing. *Renewable Energy Resources*, 2011,29(04):150-152.

[13] Zhang Jing. The application of enamel tank body on the construction of wastewater treatment plant. *Water and Wastewater Engineering*, 2006,32(10):93-96.

[14] Wei Dunman. The selection and application of technology solution of large-scale biogas. project construction. *Journal of Anhui Agricultural University*, 2013,19(18):100-102.

[15] Chen Yongsheng, Zhu Dewen, Qu Haoli. Materials pretreatment process and equipment technology of the large and medium sized biogas projects. *Chinese Agriculture Mechanization*, 2010(06): 74-78.

[16] Yan Shuiping, Chen Jinghan, Ai ping. Progress in chemical absorption technology for low-cost CO_2 removal. *Modern Chemical Industry*, 2012,32(10):25-30.

[17] Yang Mingzhen. Research on biogas engineering design for cattle farm. *Journal of Anhui Agricultural Sciences*, 2011,39(18):11072-11073.

[18] Ma Lixin, Liu Weihua, Jin Heping. The study and benefit analysis on technical equipment that used livestock manure to produce biogas. *Jiansu Agricultural Mechanization*, 2013,1:32-34.

[19] Qian Jinhua, Tian Ningning. Key problems existing in biogas project design for cattle farm China biogas. *China Biogas*, 2009,27(1):20-23.

[20] Wu Nan, Kong Chuixue, Liu Jingtao. Research progress on crop straw biogas technology China biogas. *China Biogas*, 2012,30(4):14-20.

[21] Dang Feng, Bi Yuyun, Liu Yanping. Analysis of the large-and-medium-sized biogas projects in Europe and comparisons with our country. *China Biogas*, 2014,32(1):79-89.

[22] Yang Li, Gong Naichao, Liu Langdong. A brief discussion of the current development situation on China's large and medium scale biogas plants construction. *Guide of Scitech Magazine*, 2010,36:145-146.

[23] Chen Ling, Zhao Lixin, Dong Baocheng. The status and trends of the development of biogas plants for crop straws in China. *Renewable Energy Resources*, 2010,28(3):145-148.

[24] Li Baoyu, Bi Yuyun, Gao Chunli. The current situation, problems and countermeasures of agricultural large-and-medium-scale biogas project development in China. *Journal of China Agricultural Resources and Regional Planning*, 2010,31(2):57-61.

[25] Wang Yuxin, Tang Yanfen. Design and application of large and medium-sized biogas engineering. *Beijing: Chemical Industry Press*,2013:1-163.

[26] Zhou Lixiang. Solid waste treatment disposal and recycling. Beijing: *China Agriculture Press*, 2007.

In: Gas Biofuels from Waste Biomass
Editor: Zhidan Liu

Chapter 2

BASIC CONCEPT 2: BIOHYDROGEN PRODUCTION

Xiao Wu[*]

Southern Research and Outreach Center, University of Minnesota,
Waseca, Minnesota, United States

ABSTRACT

Fossil fuel is the major source for almost all of our energy needs today, yet extensive consumption of fossil-based substances leads to problems of depleted reserve and side effects of severe environmental influences. Hydrogen is considered as "the fuel of future" mainly due to its high energy density and nonpolluting nature. Biohydrogen production is found to be less energy intensive and more environmental friendly as compared to thermochemical and electrochemical processes, while dark fermentation is the most applicable process for recovering hydrogen from wastes due to its higher rate of hydrogen evolution as well as the versatility of the substrates used. As the most abundant renewable carbon resource known with large enough capacity to substitute for fossil fuels, waste biomass is a promising substrate for biohydrogen production. This chapter presents basic concepts and fundamentals of biohydrogen production processes.

Keywords: biohydrogen, biophotolysis, fermentative hydrogen production, photo-fermentation, dark-fermentation, microbial electrolysis, hydrogen production rate, hydrogen yield

2.1. INTRODUCTION

On the energy front, both the pundits and the lay society agree that our existing energy structure that relies overwhelmingly on fossil fuels is neither sustainable nor friendly to the environment. Energy demand is expected to grow significantly in the coming decades with an increasing world population, coupled with a dwindling fossil fuel reserve (Keskin &

---

[*] Corresponding author: Southern Research and Outreach Center, University of Minnesota, 35838, 120th ST, Waseca, MN 56093, USA, Email: wuxxx199@umn.edu.

Hallenbeck, 2012b). In turn, burning fossil fuels produces greenhouse gases (GHG) that cause global climate changes that are increasingly adversely impacting the world agriculture and food production (Schmidnuber, 2007). To overcome this problem, hydrogen is suggested as a "green" energy carrier since it only produces water upon combustion. With an outstanding energy yield of 122 kJ/g, which is 2.75 times greater than hydrocarbon fuels, hydrogen is considered as an ideal fuel for the future (Tenca et al., 2011).

It has been reported that over 50 million tonnes of hydrogen are traded annually worldwide with a growth rate of nearly 10% per year since 2005 (Winter, 2005). Based on the data from the National Hydrogen Program of the United States, the contribution of hydrogen to total energy market will be 8–10% by 2025. Upon the considerably increasing demand of hydrogen, over 90% of the world's hydrogen supply is currently obtained from fossil fuels such as natural gas, petroleum, and coal. Conventional approaches for hydrogen gas production are steam reforming of methane (SRM) and other hydrocarbons (SRH), and non-catalytic partial oxidation of fossil fuels (POX) and autothermal reforming which combines SRM and POX. These are all energy intensive processes requiring high temperatures (> 850°C) (Armor, 1999). As the cleanest technology for hydrogen gas production, electrolysis of water accounts for about 4% of current hydrogen production, which requires intensive electricity that is mostly derived from fossil fuel combustion as well. Despite present technological limitations and challenges, developing sustainable methods to obtain hydrogen from renewable sources in substitution of the current fossil-fuel based technologies is paramount in order to fully achieve sustainability in both energy supply and the environment.

Biohydrogen, defined as hydrogen gas produced from renewable resources via biological ways, has drawn significant interests not only from researchers, but also from governments and energy industries around the globe due to its sustainable and economical nature. Advantages of biohydrogen production processes over the conventional methods include 1) unlimited renewable feedstocks available to use; 2) less energy intensive because they don't require high temperature or electricity; and 3) environmental friendly. Although currently possible efficiencies of biological hydrogen process are lower than that of steam methane reforming process (SMR), biohydrogen processes reduce GHG emissions and non-renewable energy use by 57-73% and 65-79%, respectively (Manish & Banerjee, 2008). Additionally, the processes are not limited to centralized energy production and well suited for small-scale installations in locations where biomass or wastes are available, thus exempting energy expenditure and transportation costs (Lo et al., 2009).

The four basic biohydrogen producing processes that have been studied are biophotolysis, photo-fermentations, dark-fermentation and microbial electrolysis cells (MECs). Among these methods, photo-fermatation and MECs decompose organic compounds, especially organic acids, while dark-fermentation degrades carbohydrates to generate hydrogen gas. Thus, various waste streams including industrial wastewater, food-processing waste, animal waste, and sludge from wastewater treatment plants can be recycled and utilized in biohydrogen production by anaerobic and photosynthetic/electrogenic microorganisms (Kapdan & Kargi, 2006), accomplishing the dual goals of waste reduction and energy generation. The potential for reliable biohydrogen production from waste biomass is colossal due to the abundance of the carbohydrate content in the biomass. However, there are also technical challenges that must be overcome before the biohydrogen production processes can be implemented at a practical or commercial level. In this chapter, the basic concept and fundamentals of biohydrogen production from waste biomass will be reviewed and limitations and challenges will also be presented.

2.2. COMPARISON OF BIOHYDROGEN PRODUCTION PROCESSES

Basic biological processes employed in hydrogen gas production can be classified in four categories by the type of microorganisms:

1. Biophotolysis of water by algae and cyanobacteria.
2. Photodecomposition of organic compounds by photo-synthetic bacteria.
3. Fermentative hydrogen production from organic compounds by anaerobic bacteria.
4. Decomposition of organic compounds by microbial electrolysis cells with electrogenic bacteria.

2.2.1. Hydrogen Procution by Biophotolysis

Biophotolysis is the most inherently appealing way to produce hydrogen by two types of microorganisms, green algae (micro-algae) and cyanobacteria, capturing solar energy and splitting water into hydrogen and oxygen. This method takes only water as substrate and uses the same processes found in the photosynthesis of plants, but generates hydrogen gas instead of carbon containing biomass. Biophotolytic processes can be divided into two types—direct and indirect (Nath & Das, 2004). In direct biophotolysis (Figure 1), two photosynthetic systems PSI and PSII are responsible for the photosynthesis process: (i) PSI producing reductant for CO_2 reduction and (ii) PSII splitting water and evolving oxygen. Two electrons that come from water can be used by PSI to realize hydrogen generation with the presence of hydrogenase, the enzyme that catalyzes the hydrogen formation by passing the reducing power of electrons to protons. Although theoretical advantages of a limitless supply of substrate and potential availability of enormous energies with a total irradiance of 1.74×10^{17} W (Hallenbeck, 2013) exist, there are serious obstacles to successful application in large capacity and practical systems. First, it requires additional engineering considerations to enclose the entire solar-energy capture area in a photo-bioreactor. Second, since one of the major products from this system is oxygen, which will strongly inhibit the hydrogenase function and hydrogen production, several steps including gas sparging need to be taken to constantly remove oxygen, which is not economically feasible (Melnicki et al., 2012). One promising solution could be to develop an oxygen-insensitive hyrdrogenase in certain types of microalgae.

On the other hand, indirect biophotolysis (Figure 2) is carried out by cyanobacteria and supposed to overcome the O_2 sensitivity in the direct biophotolysis by separating the hydrogen and oxygen evolution process into two stages. In the first stage, photosynthesis is used to capture solar energy and convert it into some form of carbohydrate stored in the cell mass of these blue-green algae, while in the second stage; carbohydrate is converted to hydrogen through fermentation (Huesemann et al., 2010). Theoretically, 12 mol of hydrogen could be produced by 1 mol glucose by this light-driven bioprocess. However, with the high cost of photobioreactors and the associated support systems needed, plus the extremely low overall light to hydrogen conversion efficiency, this approach is demonstrated to be expensive.

2.2.2. Hyrogen Production by Photo-Fermentation

Photo-fermentation is another hydrogen-producing process involving the capture and conversion of light energy, which takes place within photosynthetic bacteria (often purple non-sulfur bacteria) via the enzyme nitrogenase. Light energy and the biomass go into the single bacterial photosystem which produces two electrons and four ATP molecules. These products are then sent to the nitrogenase that is catalyzing the production of hydrogen gas. Figure 3 simply illustrates the pathways of photo-fermentation by purple non-sulfur bacteria genus *Rhodobacter*, which has been most widely studied for biohydrogen production. *Rhodobacter* species has the ability to use organic acids, simple sugars, glucose, fructose and sucrose, industrial and agricultural effluents for hydrogen production. However, they prefer to use organic acids such as acetic, butyric, propionic and lactic acids, which gives the highest conversion efficiency. Some industrial wastewater and effluent of dark-fermentation process with major content of organic acids are ideal substrates for the photo fermentation process. Having been extensively studied for 60 years, nearly stoichiometric conversion to hydrogen has been achieved by photo-fermentation (Ghosh et al., 2012a; Ghosh et al., 2012b; Sabourin-Provost & Hallenbeck, 2009). Nonetheless, low volumetric rates of hydrogen production, low-light conversion efficiencies (1-5%), and requirement for low-cost, transparent, hydrogen-impermeable photobioreactors with huge surface areas are the critical barriers for practical application of this system. Therefore, in recent research cases, photo-fermentation is often discussed as the possible second stage coupled with dark-fermentation for higher hydrogen yield (Keskin & Hallenbeck, 2012a).

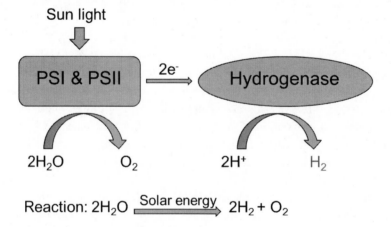

Figure 1. Illustration of direct biophotolysis pathways.

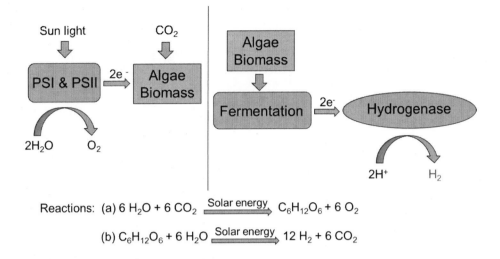

Reactions: (a) $6 H_2O + 6 CO_2 \xrightarrow{\text{Solar energy}} C_6H_{12}O_6 + 6 O_2$

(b) $C_6H_{12}O_6 + 6 H_2O \xrightarrow{\text{Solar energy}} 12 H_2 + 6 CO_2$

Figure 2. Illustration of indirect biophotolysis pathways for hydrogen production.

2.2.3. Hydrogen Production by Dark-Fermentation

Dark hydrogen fermentation is a ubiquitous step under anoxic or anaerobic conditions carried out by variety of microbes including strict anaerobes (e.g., *Clostridium sp.*) and facultative anaerobes (e.g., *Enterobacter* and *Bacillus sp.*). When these bacteria grow on organic substrates (known as heterotrophic growth), the substrates are degraded by oxidation to provide building blocks and metabolic energy for growth, where electrons are generated but note used to maintain electrical neutrality. In aerobic or oxic environments, oxygen acts as the electron acceptor with water being the product; while under anaerobic or anoxic conditions with no and/or not enough O_2 presence), protons are forced to become the electron acceptor and be reduced to molecular hydrogen (H_2) (Nandi & Sengupta, 1998). Carbohydrates, mainly glucose, are the preferred carbon sources for fermentation process, which predominantly give rise to hydrogen gas with byproduct of acetic and butyric acids (Claassen et al., 1999).

A number of metabolic pathways have potential for catabolizing sugars to hydrogen with pyruvate as the key intermediate. Figure 4 illustrates the typical biochemical pathways for conversion of glucose into hydrogen via fermentation by clostridia. In the fermentation process, pyruvate, the product of glucose catabolism, is oxidized to acetyl-CoA, which can be converted to acetyl phosphate that results in the generation of ATP and the excretion of acetate. Pyruvate oxidation to acetyl-CoA requires ferredoxin (Fd) reduction. Reduced Fd is oxidized by hydrogenase, which generates Fd and releases electrons as molecular hydrogen (Hallenbeck & Benemann, 2002). The yield of hydrogen from the breakdown of pyruvate is 2 moles from 1 mole of glucose by 2 reduced ferredoxin from pyruvate oxidation. Additional 2 moles of hydrogen can be produced by the recycle of the NAD, making theoretically a maximal hydrogen production of 4 moles by fermentation from 1 mol glucose if acetic acid is the sole byproduct.

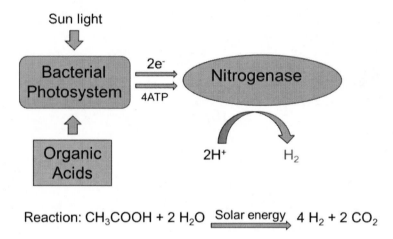

Reaction: $CH_3COOH + 2 H_2O \xrightarrow{\text{Solar energy}} 4 H_2 + 2 CO_2$

Figure 3. Illustration of photo fermentation pathways for hydrogen production.

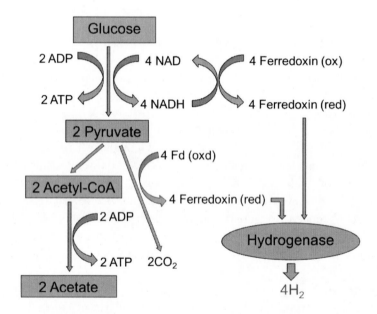

Reaction: $C_6H_{12}O_6 + 2 H_2O \rightarrow 4 H_2 + 2 CH_3COOH + 2 CO_2$

Figure 4. Illustration of dark fermentation pathways for hydrogen production.

2.2.3. Hydrogen Production by Dark-Fermentation

Dark-fermentation has been involved in the largest numbers of studies on biological hydrogen production covering a variety of different bacteria in the last two decades, due to its inherent advantages over light-dependent processes for possible industrial application. Firstly, without the requirement of light, it can produce hydrogen day and night, and the reactor technology can be much simpler as compared to that of the photo bioreactor. Although there is concern over the maintenance of anaerobic conditions when the fermentation is being carried out by strict anaerobes, e.g., clostridium, it is less of a problem when facultative anaerobic bacteria are used (Hallenbeck & Ghosh, 2009). Second, dark-fermentation can be carried out using either pure cultures or mixed cultures (e.g., bacterial consortium from anaerobic digester sludge, soil, and animal manure) that have been enriched for spore formers (strict anaerobes) by heat treatment. The advantage of mixed cultures is that they usually possess certain strength of hydrolytic activities so a variety of substrates such as lignocellulosic feedstock, organic wastes, or wastewater and complex polymeric substrates can be digested by the synergistic effect of the microflora (Hallenbeck, 2009). In addition, since anaerobic hydrogen producers have a higher growth rate to supply microorganisms to the production system, high volumetric rates of hydrogen production can be achieved. Usually, the highest hydrogen production rates are observed with immobilized mixed cultures either on inert supports or by formation of self-aggregates in the reactor over time. Granular aggregates have exhibited even greater performance over biofilm type systems (Chojnacka et al., 2011). Meanwhile, immobilized systems have been scaled up successfully to the pilot scale (Lin et al., 2011) and is the focus of industrial biohydrogen production efforts. In general, the production cost of biohydrogen by dark fermentation is 340 times lower than that of the photosynthetic process and thus is considered to be more commercially viable (Atif et al., 2005).

Despite very high hydrogen evolution rates as compared to light dependent processes, one of the main constraints of fermentative biohydrogenation process is the lower yield of hydrogen, maximally 4 mol/mol glucose, compared with the other processes (Table 1). In many cases, different pathways of fermentation with end products other than acetate, for example butyrate, which only produces 2 mol H_2/mol glucose, are more energetically favorable owing to the lower Gibbs free-energy change (shown in the formula below). Two to four mol H_2/mol glucose is the higher range of hydrogen yields associated with a mixture of acetate and butyrate fermentation, while lower hydrogen yields are with propionate and reduced endproducts (alcohols and lactic acid) (Hawkes et al., 2002). Moreover, in many organisms, the actual yields of hydrogen are further reduced by hydrogen consumption resulting from the presence of one or more hydrogen uptaking hydrogenases. In a nutshell, the yield is too low to be economically feasible as an alternative to existing chemical or electrochemical processes of hydrogen generation (Hallenbeck & Benemann, 2002). Therefore, the essential goal and challenge for fermentative hydrogen research and development need to focus significantly on obtaining higher yields of hydrogen evolution.

$$\text{glucose} + 2H_2O \rightarrow \textbf{2 acetate} + 2CO_2 + \textbf{4H}_2 \ \Delta G^{0'} = -182.4 \text{ kJ}$$
$$\text{glucose} \rightarrow \textbf{butyrate} + 2CO_2 + \textbf{2H}_2 \ \Delta G^{0'} = -257.1 \text{ kJ}$$

Considerations have to be taken into account in using dark-fermentative hydrogen producing processes to assure satisfactory productivity. First of all, to accommodate bacteria and provide them with the required conditions to achieve decent fermentative hydrogen production, bioreactor design to obtain high hydrogen evolution rates is of great importance. Different reactor designs have been examined, including batch-mode fermenter, anaerobic sequencing batch reactor (AnSBR) (Zhu et al., 2007), up flow anaerobic sludge blanket (UASB) (Jeison & Chamy, 1999), continuous stirred tank reactor (CSTR) (Wongtanet et al., 2007), granular sludge bed (Lee et al., 2006), fixed or packed bed reactor (Li et al., 2006), anaerobic fluidized bed reactor (AnFBR) (Zhang et al., 2007), and trickling bio-filter (Oh et al., 2004). The environmental and operational conditions varied in the reactors are crucial in achieving better efficiency in the anaerobic process of dark-fermentation. The conditional parameters that have been tested to be influential in the hydrogen production rate and yield included temperature, pH, hydraulic retention time (HRT), type and concentration of substrate (e.g., organic matter, nitrogen, phosphorous, and metal ions), inoculums, and organic loading rate (OLR) (Wang & Wan, 2009). Towards scaleup and commercialization of the fermentative hydrogen production, the effect of these environmental and operational factors for the chosen bioreactor type has to be investigated and optimized.

2.2.4. Hydrogen Production by Microbial Electrolysis

Hydrogen evolution from a variety of substrates by microbial electrolysis cells (MECs) is a novel technique for biohydrogen, which has seen very rapid development over the last few years (Kiely et al., 2011; Pant et al., 2012). A Microbial Electrolysis Cell (MEC) is a slight modification from MFC, where the organic matter such as organic acids in wastewater is broken down into CO_2, electrons and protons by electrochemically active microbes growing on the surface of the anode in the chamber. In MECs systems, due to the addition of voltage on the cathode in an anaerobic bioreactor, electrons and protons travel through the external circuit and solution, respectively, and combine at the cathode to generate hydrogen (Cheng & Logan, 2011). MECs have the same anodic reactions as that of MFCs, thus anode materials and microorganisms used in MFCs are also applicable in MECs. A scheme of a typical MEC is shown in Figure 5. Supplementing the voltage generated at the anode with voltage from an external power source, a MEC allows hydrogen production at the cathode from substrates whose redox potential would normally not allow it. Thus, hydrogen (−0.414 V) can theoretically be converted from acetate (−0.279 V) with the application of a relatively small voltage (−0.135 V) (Hallenbeck, 2011). In practice, additional voltage is required due to overpotentials created by physical–chemical and biological factors.

Two-chamber devices with ion-permeable membranes separating the anodic and cathodic chambers were employed in the initial research. Despite its advantage of easy capture of the separately generated hydrogen gas, the problems with this configuration include pH changes with acidification of the anode chamber due to the production of protons and alkalinization of the cathode chamber due to proton consumption. Besides, high resistance is caused by the membrane, thus increasing the voltage requirement for hydrogen evolution. These problems are soleved by single-chamber MECs, which can also generate higher current densities than dual-chamber devices and thus give significantly higher hydrogen production rates. However, the potential drawback, in turn, lies in decreasing hydrogen yields or coulombic efficiencies, plus methane production which consumes the increased hydrogen produced with this configuration, thus decreasing hydrogen yields. Another significant problem could be a biological "short circuit" using the anode as an electron sink with microbes at the anode consuming the produced hydrogen, thus creating a futile cycle and drastically increasing the amount of current needed to generate a given amount of net hydrogen (Hallenbeck, 2013).

On the economic front, the replacement of costly and rare platinum with stainless steel or various metal alloys was successful. Initial studis on effective biocathodes where hydrogen-producing microbes are immobilized on the electrode has been reported as well, although with low current densities. Further improvement in cell geometries and other physical chemical parameters may help lessen internal resistances to reduce overpotentials and the excess requirement of voltage applied. Overall, the advantages and disadvantages of basic biohydrogen production processes are summarized in Table 1.

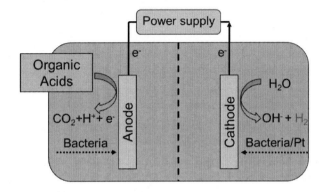

Reaction: $CH_3COOH + 2 H_2O \rightarrow 4 H_2 + 2 CO_2$

Figure 5. Illustration of typical hydrogen production by MECs.

Table 1. Comparison of four basic biohydrogen production processes

Process/ Microorganism	Substrate	Advantages	Barriers to overcome
Biophotolysis (micro-algae, cyanobacteria)	Water	Produce H_2 from water Solar energy conversion, high solar flux	Evolved O_2 incompatible with hydrogenase/nitrogenase Expensive photo bioreactors
Photo-fermentation (purple nonsulfur photosynthetic bacteria)	Organic acids Sugars Glycerol	Use wide spectrum of light Nearly complete substrate conversion Use various waste streams or effluents from dark fermenters	Low light conversion efficiencies (1-5%) Low gas production rates
Dark-fermentation (strict or facultative anaerobes)	Sugars, many different carbohydrate-rich materials	Can produce H_2 without light Can use a wide variety of waste materials as substrate Simple reactor technology High production/ growth rates	Low yields (< 4 H_2/glucose) Many side products Effluent requires further treatment before disposal CO_2 present in the biogas
Microbial electrolysis (Electrogenic bacteria)	Sugars, organic acids, proteins	Complete substrate degradation possible Produce H_2 from wide range of compounds/ organic acids	Expensive cathodes (platinum) Excess voltage needed Low current densities

2.2.5. Biohyrogen Production by Hybrid Systems

As demonstrated in the previous sections, variety of carbohydrates can be degraded by dark-fermentation at high evolution rates of hydrogen, where anaerobic bacteria decompose carbohydrates to obtain both energy and electron. Because only reactions with negative free energy could possibly proceed, organic acids formed by dark fermentation is impossible to be completely decomposed to hydrogen. In this context, hybrid two-stage systems have been suggested to increase overall energy extraction from a variety of substrates (Luo et al., 2011). After dark fermentation, the organic acid by-products can then be treated in a second stage with three possible alternatives in various stages of current development. In one scenario, the dark hydrogen fermenter is coupled with an anaerobic digestor, for which organic acids are perfect substrates. This hybrid system has been proven in several pilot scale studies to produce two gas-fuels, hydrogen and methane, as well as a significant reduction in the total chemical oxygen demand (COD). Although conventionally a single gaseous product is desirable, hydrogen and hydrogen/methane mixtures have been revealed to possess the distinguishable advantage of cleaner combustion than methane alone. Furthermore, treating the complex substrates with dark fermentation before anaerobic digestion appears to achieve greater overall substrate decomposition and higher energy recovery efficiency than single-stage methane digestion.

The other two possibilities of the second stage are the photo fermentation or MECs, where photosynthetic bacteria or electrogenic bacteria could use light energy or electric energy, respectively, to overcome the positive free energy reaction, thus produce additional hydrogen. Both of these two ways are suitable for using the organic acid effluent and, with the additional energy, converting it almost entirely to hydrogen (and carbon dioxide) (Adessi et al., 2012; Keskin & Hallenbeck, 2012a). Mixed culture of photosynthetic and anaerobic bacteria as a method of utilizing a variety of renewable resources for hydrogen production was also proposed (Yokoi et al., 1998), and the overall efficiency of hydrogen evolution can reach up to 12 mol H_2/mol hexose in the hybrid system. Both photo-fermentation and MECs processes have already been reviewed earlier, and as previously noted, will need further research before they can be applied to practice.

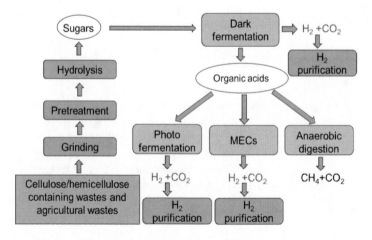

Figure 6. Schematic diagram for bio-hydrogen production from ellulose/hemicellulosic containing and agricultural waste biomass.

2.3. BIOHYDROGEN PRODUCTION FROM WASTE BIOMASS

There are various sources of biomass eligible for biofuel production: crop residues, agricultural waste, forest residues, livestock residues, energy crops, algae biomass, wood waste, food and food processing waste, biomass-based components of municipal solid waste, and industrial waste, among which the crop residues biomass that has been reported for hydrogen production includes corn stover (Cao et al., 2009), wheat straw (Kongjan & Angelidaki, 2010), wheat bran (Noike & Mizuno, 2000), rice straw (Lo et al., 2010), rice bran (Noike et al., 2005), sweet sorghum (Panagiotopoulos et al., 2010), cassava stillage (Luo et al., 2010), potato steam peels (Mars et al., 2010), beer lees (Cui et al., 2010), and sugarcane bagasse (Saratale et al., 2010). As for livestock wastes, liquid animal manure traditionally associated with biogas production is an ideal nutrient source for supplementing carbohydrate-rich wastes to produce biohydrogen (Wu et al., 2013), while dry manure such as poultry litter can be used in thermochemical conversion technologies for hydrogen production.

Given its abundance and availability, waste biomass is a promising feedstock for large-scale biohydrogen production in an economic and sustainable way (Saratale et al., 2008). However, the complex nature of these wastes may affect the biodegradability. Starch containing solid wastes is easier to process because starch can be hydrolyzed to glucose and maltose by acid or enzymatic hydrolysis followed by conversion of carbohydrates to hydrogen gas. Biomass that contains large content of cellulosic and lignocellulosic materials is more challenging to convert into biohydrogen due to the complexity in the chemical structure of such feedstock, so it must be ground and then delignified by mechanical or chemical means before fermentation. More importantly, efficient pretreatment and process design have to be contemplated in order to improve the yield and productivity of the cellulosic biohydrogen production from such waste biomass (Cheng et al., 2011; Saratale et al., 2013). Figure 6 illustrates a pathway for biohydrogen production from cellulose/hemicellulosic containing and agricultural waste biomass by anaerobic dark fermentation (first stage) followed by one of the three processes (second stage).

2.4. SUMMARY

Today, considerable research and industrial interest has been attracted to the generation of clean and sustainable energy from renewable feedstocks (such as waste biomass) for the substitution of fossil fuels. Since conventional hydrogen producing methods are all fossil fuel based and requiring great energy input, biohydrogen technology is very promising as replacement to meet the dual goal of environmental protection and sustainable energy production. Biophotolysis, photo-fermentation, dark-fermentation, and MECs are the basic biohydrogen production techniques that have been extensively studied in the past 60 years. Acting on different principles, each method has merits and demerits with respect to large scale applications. By using various types of waste resources, photo-fermentation and MECs prefer to use organic acid as substrate and they could achieve nearly perfect hydrogen yield, but the current bioreactor economics and hydrogen production rate are the limiting factors. On the other hand, dark-fermentation not only has greater ability to take all kinds of waste as feedstock, but also could obtain much higher generation rates of hydrogen gas, but with low hydrogen yield and complicity of feedstock hydrolysis being the main obstacles. Integrating photo-fermentation, MECs or anaerobic digestion with dark-fermentation could be a solution to improving overall hydrogen yield, but pretreatment of biomass and/or agricultural wastes is needed for the process.

REFERENCES

Adessi, A., De Philippis, R. & Hallenbeck, P. C. (2012). Combined systems for maximum substrate conversion. in: *Microbial Technologies in Advanced Biofuels Production*, Springer, 107-126.

Armor, J. N. (1999). The multiple roles for catalysis in the production of H2. *Applied Catalysis A: General*, *176*(2), 159-176.

Atif, A., Fakhru'l-Razi, A., Ngan, M., Morimoto, M., Iyuke, S., Veziroglu, N. (2005). Fed batch production of hydrogen from palm oil mill effluent using anaerobic microflora. *International journal of hydrogen energy*, *30*(13), 1393-1397.

Cao, G., Ren, N., Wang, A., Lee, D. J., Guo, W., Liu, B., Feng, Y. & Zhao, Q. (2009). Acid hydrolysis of corn stover for biohydrogen production using Thermoanaerobacterium thermosaccharolyticum W16. *International Journal of Hydrogen Energy*, *34*(17), 7182-7188.

Cheng, C. L., Lo, Y. C., Lee, K. S., Lee, D. J., Lin, C. Y. & Chang, J. S. (2011). Biohydrogen production from lignocellulosic feedstock. *Bioresource Technology*, *102*(18), 8514-8523.

Cheng, S. & Logan, B. E. (2011). High hydrogen production rate of microbial electrolysis cell (MEC) with reduced electrode spacing. *Bioresource Technology*, *102*(3), 3571-3574.

Chojnacka, A., Błaszczyk, M. K., Szczęsny, P., Nowak, K., Sumińska, M., Tomczyk-Żak, K., Zielenkiewicz, U. & Sikora, A. (2011). Comparative analysis of hydrogen-producing bacterial biofilms and granular sludge formed in continuous cultures of fermentative bacteria. *Bioresource technology*, *102*(21), 10057-10064.

Claassen, P., Van Lier, J., Contreras, A. L., Van Niel, E., Sijtsma, L., Stams, A., De Vries, S. & Weusthuis, R. 1999. Utilisation of biomass for the supply of energy carriers. *Applied microbiology and biotechnology*, *52*(6), 741-755.

Cui, M., Yuan, Z., Zhi, X., Wei, L. & Shen, J. (2010). Biohydrogen production from poplar leaves pretreated by different methods using anaerobic mixed bacteria. *International Journal of Hydrogen Energy*, *35*(9), 4041-4047.

Ghosh, D., Sobro, I. F. & Hallenbeck, P. C. (2012a). Optimization of the hydrogen yield from single-stage photofermentation of glucose by Rhodobacter capsulatus JP91 using response surface methodology. *Bioresource Technology*, *123*, 199-206.

Ghosh, D., Sobro, I. F. & Hallenbeck, P. C. 2012b. Stoichiometric conversion of biodiesel derived crude glycerol to hydrogen: Response surface methodology study of the effects of light intensity and crude glycerol and glutamate concentration. *Bioresource Technology*, *106*, 154-160.

Hallenbeck, P. C. (2013). Chapter 2 - Fundamentals of Biohydrogen. in: *Biohydrogen*, (Ed.) A.P.-S.C.C.H. Larroche, Elsevier. Amsterdam, 25-43.

Hallenbeck, P. C. (2009). Fermentative hydrogen production: principles, progress, and prognosis. *International Journal of Hydrogen Energy*, *34*(17), 7379-7389.

Hallenbeck, P. C. (2011). Microbial paths to renewable hydrogen production. *Biofuels*, 2(3), 285-302.

Hallenbeck, P. C. & Benemann, J. R. (2002). Biological hydrogen production: fundamentals and limiting processes. *International Journal of Hydrogen Energy*, 27(11-12), 1185-1193.

Hallenbeck, P. C. & Ghosh, D. (2009). Advances in fermentative biohydrogen production: the way forward? *Trends in Biotechnology*, 27(5), 287-297.

Hawkes, F. R., Dinsdale, R., Hawkes, D. L. & Hussy, I. (2002). Sustainable fermentative hydrogen production: challenges for process optimisation. *International Journal of Hydrogen Energy*, 27(11-12), 1339-1347.

Huesemann, M. H., Hausmann, T. S., Carter, B. M., Gerschler, J. J. & Benemann, J. R. (2010). Hydrogen generation through indirect biophotolysis in batch cultures of the nonheterocystous nitrogen-fixing cyanobacterium plectonema boryanum. *Applied Biochemistry and Biotechnology*, 162(1), 208-220.

Jeison, D. & Chamy, R. (1999). Comparison of the behaviour of expanded granular sludge bed (EGSB) and upflow anaerobic sludge blanket (UASB) reactors in dilute and concentrated wastewater treatment. *Water Science and Technology*, 40(8), 91-97.

Kapdan, I. K. & Kargi, F. (2006). Bio-hydrogen production from waste materials. *Enzyme and Microbial Technology*, 38(5), 569-582.

Keskin, T. & Hallenbeck, P. (2012a). Enhancement of Biohydrogen Production by Two-Stage Systems: Dark and Photofermentation. in: *Biomass Conversion*, (Eds.) C. Baskar, S. Baskar, R.S. Dhillon, Springer Berlin Heidelberg, 313-340.

Keskin, T. & Hallenbeck, P. C. (2012b). Hydrogen production from sugar industry wastes using single-stage photofermentation. *Bioresource Technology*, 112, 131-136.

Kiely, P. D., Regan, J. M. & Logan, B. E. (2011). The electric picnic: synergistic requirements for exoelectrogenic microbial communities. *Current opinion in biotechnology*, 22(3), 378-385.

Kongjan, P. & Angelidaki, I. (2010). Extreme thermophilic biohydrogen production from wheat straw hydrolysate using mixed culture fermentation: Effect of reactor configuration. *Bioresource Technology*, 101(20), 7789-7796.

Lee, K-S., Lo, Y-C., Lin, P-J. & Chang, J-S. (2006). Improving biohydrogen production in a carrier-induced granular sludge bed by altering physical configuration and agitation pattern of the bioreactor. *International Journal of Hydrogen Energy*, 31(12), 1648-1657.

Li, C., Zhang, T. & Fang, H. H. P. (2006). Fermentative hydrogen production in packed-bed and packaging-free upflow reactors. in: *Water Science and Technology*, Vol. 54, 95-103.

Lin, C-Y., Wu, S-Y., Lin, P-J., Chang, J-S., Hung, C-H., Lee, K-S., Lay, C-H., Chu, C-Y., Cheng, C-H. & Chang, A. C. (2011). A pilot-scale high-rate biohydrogen production system with mixed microflora. *international journal of hydrogen energy*, 36(14), 8758-8764.

Lo, Y-C., Saratale, G. D., Chen, W-M., Bai, M-D. & Chang, J-S. (2009). Isolation of cellulose-hydrolytic bacteria and applications of the cellulolytic enzymes for cellulosic biohydrogen production. *Enzyme and Microbial Technology*, 44(6–7), 417-425.

Lo, Y. C., Chen, C. Y., Lee, C. M. & Chang, J. S. (2010). Sequential dark-photo fermentation and autotrophic microalgal growth for high-yield and CO2-free biohydrogen production. *International Journal of Hydrogen Energy*, 35(20), 10944-10953.

Luo, G., Xie, L., Zhou, Q. & Angelidaki, I. (2011). Enhancement of bioenergy production from organic wastes by two-stage anaerobic hydrogen and methane production process. *Bioresource technology*, *102*(18), 8700-8706.

Luo, G., Xie, L., Zou, Z., Zhou, Q. & Wang, J. Y. (2010). Fermentative hydrogen production from cassava stillage by mixed anaerobic microflora: Effects of temperature and pH. *Applied Energy*, *87*(12), 3710-3717.

Manish, S. & Banerjee, R. (2008). Comparison of biohydrogen production processes. *International Journal of Hydrogen Energy*, *33*(1), 279-286.

Mars, A. E., Veuskens, T., Budde, M. A. W., Van Doeveren, P. F. N. M., Lips, S. J., Bakker, R. R., De Vrije, T. & Claassen, P. A. M. (2010). Biohydrogen production from untreated and hydrolyzed potato steam peels by the extreme thermophiles Caldicellulosiruptor saccharolyticus and Thermotoga neapolitana. *International Journal of Hydrogen Energy*, *35*(15), 7730-7737.

Melnicki, M. R., Pinchuk, G. E., Hill, E. A., Kucek, L. A. & Fredrickson, J. K., Konopka, A. & Beliaev, A. S. (2012). Sustained H2 production driven by photosynthetic water splitting in a unicellular cyanobacterium. *mBio*, *3*(4).

Nandi, R. & Sengupta, S. (1998). Microbial production of hydrogen- An overview. *Critical Reviews in Microbiology*, *24*, 61-84.

Nath, K. & Das, D. (2004). Improvement of fermentative hydrogen production: various approaches. *Applied Microbiology and Biotechnology*, *65*(5), 520-529.

Noike, T., Ko, I., Yokoyama, S., Kohno, Y. & Li, Y. (2005). Continuous hydrogen production from organic waste. *Water Science & Technology*, *52*(1-2), 145-151.

Noike, T. & Mizuno, O. (2000). Hydrogen fermentation of organic municipal wastes. *Water Science and Technology*, *42*(12), 155-162.

Oh, Y. K., Kim, S. H., Kim, M. S. & Park, S. (2004). Thermophilic biohydrogen production from glucose with trickling biofilter. *Biotechnology and bioengineering*, *88*(6), 690-698.

Panagiotopoulos, I. A., Bakker, R. R., De Vrije, T., Koukios, E. G. & Claassen, P. A. M. (2010). Pretreatment of sweet sorghum bagasse for hydrogen production by Caldicellulosiruptor saccharolyticus. *International Journal of Hydrogen Energy*, *35*(15), 7738-7747.

Pant, D., Singh, A., Van Bogaert, G., Olsen, S. I., Nigam, P. S., Diels, L. & Vanbroekhoven, K. (2012). Bioelectrochemical systems (BES) for sustainable energy production and product recovery from organic wastes and industrial wastewaters. *Rsc Advances*, *2*(4), 1248-1263.

Sabourin-Provost, G. & Hallenbeck, P. C. (2009). High yield conversion of a crude glycerol fraction from biodiesel production to hydrogen by photofermentation. *Bioresource Technology*, *100*(14), 3513-3517.

Saratale, G., Saratale, R. & Chang, J-S. (2013). Biohydrogen from renewable resources. *Biohydrogen. Elsevier Plc*, 185-221.

Saratale, G .D., Chen, S. D., Lo, Y. C., Saratale, R. G. & Chang, J. S. (2008). Outlook of biohydrogen production from lignocellulosic feedstock using dark fermentation - A review. *Journal of Scientific and Industrial Research*, *67*(11), 962-979.

Saratale, G. D., Saratale, R. G., Lo, Y. C. & Chang, J. S. (2010). Multicomponent cellulase production by Cellulomonas biazotea NCIM-2550 and its applications for cellulosic biohydrogen production. *Biotechnology progress*, *26*(2), 406-416.

Schmidnuber, J. & Tubiello, F. N. (2007). Global food security under climate change. in: *Proceedings of the National Academy of Sciences of the United States of America*, (Ed.) W. Easterling, 19703-19708.

Tenca, A., Schievano, A., Perazzolo, F., Adani, F. & Oberti, R. (2011). Biohydrogen from thermophilic co-fermentation of swine manure with fruit and vegetable waste: Maximizing stable production without pH control. *Bioresource Technology*, *102*(18), 8582-8588.

Wang, J. & Wan, W. (2009). Factors influencing fermentative hydrogen production: A review. *International Journal of Hydrogen Energy*, *34*(2), 799-811.

Winter, C-J. (2005). Into the hydrogen energy economy—milestones. *International Journal of Hydrogen Energy*, *30*(7), 681-685.

Wongtanet, J., Sang, B. I., Lee, S. M. & Pak, D. (2007). Biohydrogen production by fermentative process in continuous stirred-tank reactor. *International Journal of Green Energy*, *4*(4), 385-395.

Wu, X., Lin, H. & Zhu, J. (2013). Optimization of continuous hydrogen production from co-fermenting molasses with liquid swine manure in an anaerobic sequencing batch reactor. *Bioresource Technology*, *136*, 351-359.

Yokoi, H., Mori, S., Hirose, J., Hayashi, S. & Takasaki, Y. (1998). H2 production from starch by a mixed culture of Clostridium butyricum and Rhodobacter sp. M [h] 19. *Biotechnology letters*, *20*(9), 895-899.

Zhang, Z.-P., Tay, J.-H., Show, K.-Y., Yan, R., Liang, D. T. & Lee, D.-J., Jiang, W.-J. (2007). Biohydrogen production in a granular activated carbon anaerobic fluidized bed reactor. *International Journal of Hydrogen Energy*, *32*(2), 185-191.

Zhu, J., Wu, X., Miller, C., Yu, F., Chen, P. & Ruan, R. (2007). Biohydrogen production through fermentation using liquid swine manure as substrate. *Journal of Environmental Science and Health Part B*, *42*, 393-401.

PART II. TECHNICAL ADVANCES ON GAS BIOFUELS PRODUCTION

In: Gas Biofuels from Waste Biomass
Editor: Zhidan Liu

ISBN: 978-1-63483-192-5
© 2015 Nova Science Publishers, Inc.

Chapter 3

PRETREATMENT OF COMPLEX BIOMASS FOR BIOFUELS

Laiqing Li and Mingxia Zheng[*]

Department of Environmental Engineering,
Beijing University of Chemical Technology, Beijing, China

ABSTRACT

Lignocellulosic biomass is an abundant, non food-based, renewable, and low-cost resource for the production of renewable fuels and chemicals. However, its inherent recalcitrance to biological conversion hinders its application for commercial production of biofuels. The goal of the pretreatment process is to release cellulose and hemicellulose from the lignin seal and its crystalline structure in order to render polysaccharides accessible for microorganisms. For a long time, a number of pretreatment methods have been developed and applied for mainly efficient conversion of (hemi-) cellulose to ethanol, methane and, in the last years, also to hydrogen. In this chapter, the published investigations on pretreatment methods priory to biofuel conversion were reviewed. For each pretreatment method, the process, reactions, and changes incorporated in the biomass during pretreatment as well as the efficiencies, advantages, and drawbacks are presented. Furthermore, the recent advances of the technology, future perspective and challenges encountered throughout the entire process are also reviewed.

Keywords: lignocellulosic biomass, pretreatment, biofuel

INTRODUCTION

Lignocellulosic biomass, the source for biofuels, is abundant, sustainable, and low-cost for the production of renewable fuels and chemicals. Growing demand for human food and considering the priority for starving human society could make the lignocellulosic materials

---

[*] Corresponding author: Email: zhengmingxia@gmail.com.

potentially competitive and perhaps cost-effective feedstocks in the near future compared to sugar and starch based materials such as sugarcane and grains (Taherzadeh and Karimi, 2007; Talebnia et al., 2010). However, the drawback of lignocellulosic materials is their recalcitrance to biological conversion, which increases the manufacturing cost of the biofuels. The contents of cellulose, hemicellulose, and lignin in common residues are listed in Table 3.1.

Pretreatment is the process used to release cellulose and hemicellulose from the lignin seal and its crystalline structure so as to render polysaccharides accessible for microorganisms. For a long time, many researches are being done to enhance the digestibility of lignocellulosic biomass for mainly the efficient conversion of (hemi-) cellulose to ethanol, methane and, in the last years, also to hydrogen. A number of pretreatment methods have been developed and applied for lignocellulosic biomass. The pretreatments are roughly classified into physical, physico-chemical, chemical and biological processes (Figure 3.1).

Table 3.1. The content of cellulose, hemicellulose, and lignin in common residues and wastes[a]

Lignocellulosic materials	Cellulose (%)	Hemicellulose (%)	Lignin (%)
Primary wastewater solids	8-15	NA[b]	24-29
Swine waste	6.0	28	NA
Solid cattle manure	1.6-4.7	1.4-3.3	2.7-5.7
Hardwoods stems	40-55	24-40	18-25
Softwood stems	45-50	25-35	25-35
Nut shells	25-30	25-30	30-40
Corn cobs	45	35	15
Corn stover	34–36	22–29	7–20.2
Wheat straw	33–50	24–36	9–17
Sugarcane bagasse	40	24	25
Rice straw	28–47	19–25	10–25
Grasses	25-40	35-50	10-30
Office paper	68.6	12.4	11.3
Wheat straw	30	50	15
Sorted refuse	60	20	20
Leaves	15-20	80-85	0
Cotton seed hairs	80-95	5-20	0
Cotton stalk	31	11	28
Newspaper	40-55	25-40	18-30
Waste papers from chemical pulps	60-70	10-20	5-10
Coastal Bermuda grass	25	35.7	6.4
Swith grass	45	31.4	12.0

[a]Source: Reshamwala et al., (1995), Sun and Cheng (2002), Jouanin and Lapierre (2012), Hu et al., (2012), Shafiei et al., (2015).
[b]NA – not available.

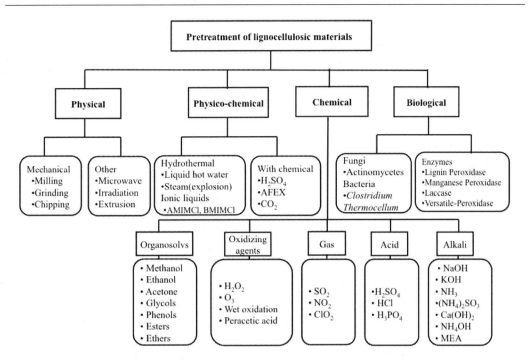

Figure 3.1. Schematic overview of different methods for pretreatment of lignocellulosic materials.

The applied methods usually use combination of different principles, such as mechanical together with thermal and chemical effects in order to achieve high sugar release efficiencies, low toxicants production, and low energy consumption. The overall efficiency of the pretreatment process is correlated to a good balance between low inhibitors formation and high substrate digestibility. Pretreatment must meet the following requirements (Sun and Cheng, 2002; Rajendran and Taherzaden, 2014; Shafiei et al., 2015):

(1) **Efficiency of pretreatment** : achievement of high saccharification yields in subsequent enzymatic hydrolysis.

(2) **Suitability for different types of feedstocks:** as most of lignocelluloses feedstocks are mixtures of different types of materials, and different feedstocks may be with different moisture content.

(3) **Materials recovery:** minimum the degradation or loss of carbohydrates;

(4) **Inhibitory avoid**: minimum the formation of byproducts inhibitory to the subsequent hydrolysis and fermentation processes;

(5) **Energy consumption**: most of the pretreatments are for processes where energy is the final product, such as ethanol, butanol, or biogas. Consequently, the energy consumption of the pretreatment is very important to the energy balance of the whole process.

(6) **Cost-effective:** minimal capital for pretreatment and auxiliary (e.g., neutralization and washing) equipment; minimal or no chemical for pretreatment;   minimal utility (e.g., steam, cooling water, and electricity) requirements

(7) **Risks and environmental aspects**: minimum production of waste chemicals and no discharge of toxic or   hazardous wastes

Although no pretreatment meets all the specifications stated above, these aspects should be considered for an economically feasible process. The objective of this chapter is to review the published investigations on pretreatment methods priory to biofuel conversion and to present the recent advances of the technology, future perspective and challenges encountered throughout the entire process. Moreover, additional problems will be analyzed, and finally, conclusions with respect to promising pretreatment techniques and needed future research are made.

3.1. PHYSICAL PRETREATMENT

Physical pretreatments involve changes in the structure of biomass without addition of chemicals or production of harmful residues. The general objective of this pretreatment category is to reduce particle size and crystallinity of lignocelluloses, thus increasing surface area accessibility, and decreasing degree of polymerization (DP) (Alvira et al., 2010; Rajendran and Taherzaden, 2014). Mechanical comminution such as chipping, grinding or milling, extrusion, pyrolysis, irradiation, ultrasonication, and using microwaves are among the physical pretreatment methods (Taherzadeh and Karimi, 2008; Agbor et al., 2014; Shafiei et al., 2015). Each of the method is suitable for raw materials with specific properties.

3.1.1. Mechanical Pretreatment

The objective of a mechanical pretreatment is a reduction of particle size and crystallinity. Biomass materials can be comminuted by a combination of chipping, grinding and milling. The reduction in particle size leads to an increase of available specific surface and reductions of cellulose crystallinity and the degree of DP (Palmowski and Muller, 1999). The size of the materials is usually 10–30 mm after chipping and 0.2–2 mm after milling or grinding. Vibratory ball milling has been found to be more effective in breaking down the cellulose crystallinity of spruce and aspen chips and improving the digestibility of the biomass than ordinary ball milling (Millet et al., 1976). Milling does not change the composition of cellulose and hemicelluloses or the lignin content. The increase in specific surface area, reduction of DP, and the shearing, are all factors that increase the total hydrolysis yield of the lignocellulose in most cases by 5–25% (depends on kind of biomass, kind of milling, and duration of the milling), but also reduces the technical digestion time by 23–59% (thus an increase in hydrolysis rate) (Delgenés et al., 2002; Hendriks and Zeeman, 2009). Milling causes both an increased methane (5–25%; Delgenés et al., 2002) and ethanol yield and also increases the hydrolysis rate. However, use of very small particles may not be desirable due to higher energy consumption in milling stage as well as imposing negative effect on the subsequent pretreatment method. A particle size reduction below 40 mesh has little effect on the hydrolysis yield as well as hydrolysis rate of the biomass as reported by Chang and Holtzapple (2000). Energy consumption is a very important factor in physical pretreatment more so than in other pretreatments. Initial and ultimate particle size, moisture content and material properties are among variables that influence both energy consumption and the effectiveness of subsequent processing. A comparison of energy requirement of mechanical comminution is shown in Table 3.2. Smaller particle size and higher moisture content of materials will lead to higher specific energy consumption.

Different milling processes are used based on the material and its nature. For instance, colloid mill, fibrillator, and dissolver are compatible with wet materials, while extruder, roller mill, cryogenic mill, and hammer mill are designed for dry materials. Ball milling can be used for both dry and wet materials (Walpot, 1986; Rajendran and Taherzaden, 2014). As no production of inhibitors (like furfural and hydroxym-ethylfurfural (HMF)) are produced, mechanical comminution is suited for both methane and ethanol production.

3.1.2. Extrusion

Extrusion is a novel pretreatment method applied to wet biomass containing over 15–20% moisture content (Zheng et al., 2014) and is a combination of heating, shearing, and mixing. The combination of these different actions results in the lignocelluloses being physically as well as chemically modified. The mechanism of extrusion is that high mechanical shear disrupts the biomass structure and leads to defibrillation and shortening of the fibers.

There are two kinds of extruders, i.e., single screw extruders and twin screw extruders. Single screw extruders consist of a simple screw rotating in a barrel. This type of extruder is used for spatial mixing of materials and is not efficient for changing physical properties of the materials by application of high shear. The efficient enhancement of enzymatic hydrolysis cannot be achieved by application of sole extrusion; however, this method has been used in continuous pretreatments in combination with other physical or chemical pretreatment methods. Ongoing research on pretreatment methods is exploring the potential of combining enzymatic hydrolysis and chemicals such as NaOH (Vandenbossche et al., 2014; Duque et al. 2014; Zhang et al., 2012), dilute acid (Wang et al., 2014; Ciesielski et al., 2014), Na_2S, and H_2O_2 with extrusion (Choi and Oh, 2012; Shafiei et al., 2015).

Twin screw extruders are more efficient than the single ones. Extruders with fully intermeshing screws with co- or counter-rotating screws are useful for generation of high shear. For instance, regular flighted right- or left- handed and kneading disk screw types have been suggested for pretreatment (Kalyon and Malik, 2007). Changing the physical properties is possible when the whole fluid can be forced to the tight clearance between screws or screw and barrel. This means that the separation of solid and liquid should not occur otherwise the high shear is not applied to the solid (Senturk-Ozer et al., 2011; Shafiei et al., 2015).

Compare to mechanical comminution, the energy required for extrusion is low. Another advantage of extrusion is application of high shear and rapid mixing beside application of steam and addition and removal of chemicals. Main problem with the extrusion of lignocelluloses is limitations in flowability of materials. This problem leads to separation of liquid from solid in extruder, and thus high shearing stresses are not applied to biomass. Therefore, the addition of chemicals was suggested for enhancement of flowability behavior of biomass. Chemicals such as carboxy methyl cellulose in combination with NaOH or recycled black liquor were found to be efficient for enhancement of flowability of the solids

(Senturk-Ozer et al., 2011; Shafiei et al., 2015). But this may increase the environmental problems.

3.1.3. Irradiation

Irradiation is performed by means of gamma radiation, ultrasound, electron beam, and microwaves (Zhao et al., 2012). The cellulose fraction of lignocelluloses is degraded by irradiation to breakable fibers, low molecular oligosaccharides, or even to cellobiose (Kumakura and Kaetsu, 1983). Therefore, the enzymatic degradation of cellulose to glucose can be enhanced in sequence. The irradiation higher than 100 MR can result in glucose and small oligosaccharides being decomposed into low molecular compound. Lignin is removed only to a limited extent by irradiation. Notwithstanding, the described methods are still expensive and are facing challenges in the application on an industrial scale (Kumakura and Kaetsu, 1983, 1984; Rajendran and Taherzaden, 2014).

Ultrasound is one of the irradiation pretreatment types and its disintegration of lignocelluloses is dependent on the density and intensity of the ultrasonic waves (McDermott et al., 2001). This pretreatment might be the irradiation method closest to commercialization; ultrasounds for improving biogas production from grasses are now available in the market. (Alvira et al., 2010; Rajendran and Taherzaden, 2014). Ultrasound has also been used in combination with other physical or chemical pretreatment methods, such as cholinium ionic liquids (Ninomiya et al., 2013). Ninomiya et al. (2013) demonstrated that, for the first time, that the cholinium-IL-assisted pretreatment of lignocellulosic biomass is enhanced by ultrasound irradiation in comparison with conventional heating. The cellulose saccharification ratio of bamboo powder was approximately 55% when pretreated thermally in choline acetate at 110 °C for 60 min.

Table 3.2. Energy requirement of mechanical comminution of agricultural lignocellulosic materials with different size reduction (Cadoche and López, 1989; Sun and Cheng, 2002)

Lignocellulosic materials	Final size (mm)	Energy consumption (kWh/ton)	
		Knife mill	Hammer mill
Hardwood	1.60	130	130
	2.54	80	120
	3.2	50	115
	6.35	25	95
Straw	1.60	7.5	42
	2.54	6.4	29
Corn stover	1.60	NA[a]	14
	3.20	20	9.6
	6.35	15	NA[a]
	9.5	3.2	NA[a]

[a] NA – not available.

Microwave pretreatment has thermal as well as nonthermal effects and is carried out by immersing biomass in dilute chemical agents, subsequently exposing it to microwaves for 5–20 minutes (Alvira et al., 2010; Rajendran and Taherzaden, 2014). Microwave treatment has a high heating efficiency and is easy to implement. Studies have shown that microwave irradiation changes the ultrastructure of cellulose (Xiong et al., 2000), degrades lignin and hemicellulose in lignocellulosic materials, and increases the enzymatic susceptibility of lignocellulosic materials (Azuma et al., 1984; Ooshima et al., 1984). Microwaves have been used in pretreatment of rice straw, wheat straw and switchgrass and ethanol yields comparable to those from conventional pretreatment methods were achieved (Hu and Wen, 2008; Ma et al., 2009; Lu et al., 2011). However, most investigations have employed bench-scale batch reactors, which are only suitable for pretreating small amounts of biomass. Peng et al., (2014) developed a continuous microwave irradiation reactor increasing the scale of the microwave irradiation-assisted pretreatment technology. Before implementation of microwave treatments, it is necessary to ascertain that the energy consumption of this process is low enough to achieve economical energy production. The energy efficiency of microwave-assisted dilute sulfuric acid pretreatment of rape straw for the production of ethanol was investigated by Lu et al., (2011). They found that in general, 1 g ethanol could produce about 30 kJ of energy, and therefore, the energy input for the pretreatment was only 35% of the energy output.

Application of gama irradiation electron beam, ultrasonication, and microwaves is mostly suitable for wet feedstocks. However, these methods are very costly and require safety concerns (Karimi et al. 2013; Shafiei et al., 2015).

3.2. PHYSICO-CHEMICAL PRETREATMENT

Physicochemical pretreatments exploit and use conditions, solvent systems or compounds that affect the physical and chemical properties of biomass. The solubilization of lignocellulose components depends on pH, temperature, and moisture content. In lignocellulosic materials such as corn stover, hemicellulose is the most thermal-chemically sensitive fraction. Hemicellulose compounds start to solubilize into the water at temperature > 150°C and among various components, xylan can be extracted the most easily (Hendriks and Zeeman, 2009). Therefore, physicochemical pretreatments are the most efficient and widespread forms of pretreatments, which includes the popular pretreatment technologies such as steam explosion/steam pretreatment (ST/SE), liquid hot water pretreatment (LHW), ammonia fiber/freeze explosion (AFEX), CO_2 explosion, and ionic liquid-based pretreatment. In the following sections some of these common pretreatment methods are presented.

3.2.1. Steam Pretreatment/ Steam Explosion (ST/SE)

(1) General Steam Pretreatment/Steam Explosion

Steam pretreatment/steam explosion is one of the most widely used methods for lignocellulosic biomass pretreatment (Ballesteros et al., 2006). In this method, size-reduced biomass is put in a large vessel and rapidly heated by high pressure steam for a period of time. After a set time, the steam is suddenly released and the biomass is quickly cooled down which makes the materials undergo an explosive decompression. The difference between 'steam' pretreatment and 'steam explosion' pretreatment is the quick depressurization and cooling down of the biomass at the end of the steam explosion pretreatment, which causes the water in the biomass to 'explode' (Hendriks and Zeeman, 2009). In this section steam explosion will be mainly discussed. Steam explosion is typically initiated at a temperature of 160–260 °C (corresponding pressure 0.69–4.83 MPa) for several seconds to a few minutes before the material is exposed to atmospheric pressure (Sun and Cheng, 2002). It has been studied for several raw materials. The efficiency of this pretreatment is higher for agricultural residues and hardwoods than the efficiency for softwoods. This process has been demonstrated in several pilot plants in batch or continues modes of operation, pilot plants are presently operating in Sweden and Canada (Galbe and Zacchi, 2007) and therefore commercial equipment is available for this pretreatment (Elander, 2013).

In a steam explosion process with no chemical added to the biomass, parts of the hemicellulose hydrolyze and form kinds of acids, which could further catalyze the hydrolysis of the hemicellulose. This process, in which the in situ formed acids catalyze the process itself, is called 'auto-cleave' steam pretreatment (Hendriks and Zeeman, 2009). However, the role of the acids is probably not to catalyze the solubilization of the hemicellulose, but to catalyze the hydrolysis of the soluble hemicellulose oligomers (Hendriks and Zeeman, 2009). Mechanical forces are applied when the pressure of media is suddenly released. Water is trapped inside biomass, and a portion of water is changed to vapor and suddenly expands as pressure is decreased. This process disrupts biomass materials to smaller parts and increases accessible surface areas for hydrolytic enzymes. The result of steam explosion is explosive decompression of biomass (Zheng et al., 2014), decrease the particle sizes, separation of fibers, partial solubilization of hemicelluloses, removal and/or redistribution of lignin, and changes in cellulose structure (Brodeur et al., 2011). All these changes make the biomass more digestible (Zheng et al., 2014; Shafiei et al., 2015).

Three products are derived from steam explosion pretreatment: (1) a solid containing less recalcitrant cellulose plus lignin, (2) a liquid containing solubilized hemicelluloses plus some degradation products of lignin and pentoses, e.g., furan derivatives and phenolic compounds, and (3) water vapor containing volatile compounds which are produced during pretreatment, e.g., 60–70% of the produced furfural (Aden et al., 2002; Shafiei et al., 2015). The amount of liquid fraction depends on the moisture content of substrate, which directly affects the amount of steam required for heating the media., The liquid fraction contains 40–90% (depending on the pretreatment conditions and type of biomass) of biomass hemicellulose. Fermentation of hemicelluloses is economically feasible when the sugar concentration is over 10% (Taherzadeh and Karimi 2008; Karimi et al., 2013). The solid product from steam explosion can be washed to remove toxic compounds. However, inhibitors are still present in the liquid fraction. Detoxification methods are necessary prior to hydrolysis and fermentation (Karimi et al., 2013; Shafiei et al., 2015).

The factors that affect steam explosion pretreatment are temperature, residence time, particle size and moisture content. The process severity is a function of pretreatment time and temperature. The moisture content of the biomass influences the needed pretreatment time. The higher the moisture content, the longer the optimum steam pretreatment times (Hendriks and Zeeman, 2009). The optimum temperature highly depends on biomass type and is in the range of 180–220 °C, whereas the optimum time is 2–8 min (Taherzadeh and Karimi, 2008).

Beltrame et al. (1992) studied the effect of steam explosion on the fractionation of wheat straw. The maximum delignification occurred with pretreatment at 210 °C and 1–2 min whereas the maximum solubilization of cellulose-rich solid fraction (83.7%) and the highest glucose production (93.5%) during enzymatic hydrolysis attained at 230 °C and 1 min..

A common term used in steam pretreatment is the so called 'severity factor' (log R_0), which is a measure for the severity of the pretreatment. In this severity factor the temperature of the pretreatment and the duration of the pretreatment are combined in the following way (Overend and Chornet, 1987):

$$\log R_0 = \log(t^* \, e^{((T\square-100)/14.75)}) \qquad (3.1)$$

where with 't' in minutes and 'T' in degrees Celsius. Higher severity, e.g., higher temperature and residence time, results in more solubilization of hemicellulose and better saccharification of cellulose but simultaneously promotes the production of more inhibitor compounds. Thus, the pretreatment process must be optimized to achieve optimum yields for glucose and other sugars along with minimum inhibitory effect. Linde et al., (2008) investigated less severe condition in impregnation step in a study. The highest overall yield of the sum of glucose and xylose was obtained with treatment at 190 °C and 10 min for the wheat straw. This result is uncommon for other lignocellulosic materials to reach the maximum yield of hemicellulosic and cellulosic sugars at the same pretreatment conditions.

The advantages of steam explosion pretreatment without addition of chemicals include the requirement of relatively low capital investment, low environmental impacts, high carbohydrate recovery, moderate energy requirements compared to mechanical comminution and processes with less hazardous conditions and materials (Avellar and Glasser 1998). The conventional mechanical methods require 70% more energy than steam explosion to achieve the same size reduction (Sun and Cheng, 2002).

Limitations of steam explosion include low saccharification yields, incomplete disruption of the lignin–carbohydrate matrix, and losses of carbohydrates due to solubilization along with formation of inhibitory compounds to microorganisms used in downstream processes (Taherzadeh and Karimi, 2008). Because of the formation of degradation products that are inhibitory to the following hydrolysis and fermentation process, pretreated biomass needs to be washed by water to remove the inhibitory compounds along with water-soluble hemicellulose (McMillan, 1994). However, the water wash decreases the overall saccharification yields due to the removal of soluble sugars, such as those generated by hydrolysis of hemicellulose. Typically, 20–25% of the initial dry matter is removed by water wash (Mes-Hartree et al., 1988). The efficiency of single steam explosion as the sole pretreatment method is not enough high. Higher yields are only obtained when special additions (e.g., acid, base, or CO_2) or special types of pretreatment reactors (e.g., percolation or extrusion reactor) are applied. A single-step steam explosion at high severity results in more digestible cellulose and simultaneously increases the degradation of sugars to inhibitors. This problem can also been overcome by application of two-step steam explosion. .

(2) Thermal Pretreatment in Combination with Acid Pretreatment

Application of steam explosion in combination with acid has been widely studied over the years for ethanol production prior to concentrated acid hydrolysis (Mosier et al., 2005).

SO_2 is more commonly used for softwood compared with sulfuric acid (Zhu and Pan, 2010). The pretreatment process includes mixing the substrate with sulfuric acid in liquid form or SO_2 in vapor phase, steam heating, and explosive discharge of the materials. Typically, pretreatment is performed at 190–210 °C for hardwood and 200–220 °C for softwood with 0–5% acid or SO_2 added to biomass (Zhu and Pan, 2010). During steam pretreatment the SO_2 is converted to H_2SO_4 in the first 20 seconds of the process; after that, the catalytic hydrolysation of the hemicellulose starts. The basic reactions in acid-catalyzed steam explosion are similar to acid-catalyzed pretreatment. This effect of addition of an external acid is to catalyze the solubilization of the hemicellulose, lower the optimal pretreatment temperature and give a better enzymatic hydrolysable substrate (Gregg and Saddler, 1996; Hendriks and Zeeman, 2009). Ballesteros et al., (2006) investigated the effect of steam explosion on the diluted acid (0.9%) or water-impregnated wheat straw in various temperatures and residence times, and they found the best results obtained from steam explosion of acid-impregnated wheat straw at 180 °C and 10 min.

The explosive discharge of materials at the end of process disrupts the biomass structure to lower size of particles, and thus increases its specific surface area (McMillan, 1992; Brownell et al., 1986; Ballesteros et al., 2000). Physical disruption of material after acid pretreatment should be more efficient compared with un-catalyzed steam explosion, due to higher efficiency of the pretreatment. This advantage seems most beneficial in industrial scale processes where energy and equipment costs for size reduction are considerable. In some acid-catalyzed pretreatments, materials are cooled by a coolant media such as water or nitrogen quenching. These methods are considered not economically feasible in industrial scale processes. Discharge of the materials to a blow down vessel with ambient pressure facilitates rapid temperature reduction to boiling point of water. Furthermore, the explosive discharge facilitates fast emptying of the pretreatment reactor. This is equivalent to less residence time of material in the reactor and smaller pretreatment equipment. However, this pretreatment needs additional costs of the blowdown vessel; moreover, the vapor product is highly corrosive and must be cooled and treated in the wastewater treatment.

Some studies suggest that the application of acid increases the inhibitor compounds (Shafiei et al., 2015). Grohmann et al., (1985) reported that an appreciable production of furfural occured during steam pretreatment at temperatures of 160 °C and higher with 0.5% sulfuric acid addition. Tengborg et al., (1998) also showed a severe inhibition in the ethanol production step at a severity factor of 3 and with the addition of an external acid. When similar temperature and residence time are used for acid-catalyzed steam explosion and un-catalyzed steam explosion, more inhibitor compounds are produced in the former one. However, in such conditions, higher glucose yields are obtained with the addition of acid. The vapor product can remove a part of inhibitor compounds and since a part of water is evaporated, higher xylose concentrations are obtained compared with simple dilute acid pretreatment which is advantageous for the subsequent hydrolysis and fermentation processes (Zhu and Pan, 2010; Shafiei et al., 2015). Dilute acid steam explosion is mostly used for pretreatment of woody biomass. High efficiencies were obtained by acid-catalyzed steam explosion for hardwood (> 80%); however, the results were not sufficiently high for softwood (< 70%) (Shafiei et al., 2015). Several possibilities for operating acid-catalyzed explosion have been studied in laboratory or pilot scale experiments. The equipment for continuous pretreatment in commercial scales is also available (Aden et al., 2002; Aden and Foust, 2009; Elander, 2013).

(3) Thermal Pretreatment in Combination with Alkaline Pretreatment

Besides acid-catalyzed steam explosion, another way to improve the thermal pretreatment is to add an external alkali instead of an acid to the steam explosion process. Lime, NaOH, and $Ca(OH)_2$ is the most used external alkali. This pretreatment is usually carried out at temperatures of 100–150 °C with lime addition of approximately 0.1 g $Ca(OH)_2$ g substrate^{-1} (Chang et al., 2001a, b). Significant enhancements in ethanol and glucose yields were achieved by these methods (Sun et al., 2014; Wanderley et al., 2013). According to Kaar and Holtzapple (2000), lime pretreatment (with heating) is sufficient to increase the digestibility of low-lignin containing biomass, but not for high-lignin containing biomass. Chang and Holtzapple (2000) attributed the effectiveness of lime pretreatment to the opening of the 'acetyl valve' and partly opening the 'lignin valve', making the substrate more accessible to hydrolysis. A two-step pretreatment combined with steam explosion and alkaline for ethanol production from sugarcane bagasse was studied by Oliveria et al., (2013). Steam explosion was performed in the first step and then followed by alkaline delignification. The overall process efficiently removed about 90% of hemicellulose and lignin and obtained 80% glucose yield from the solid fraction. However, total cellulose losses during pretreatments were 37–43% which is a major drawback for this pretreatment. A positive effect of lime is that it is relatively cheap and safe (Gandi et al., 1997) and that the calcium can be regained as insoluble calcium carbonate by the reaction with carbon dioxide. This calcium carbonate can be converted to lime again with the lime kiln technology (Chang et al., 1998).

3.2.2. Ammonia Fiber Explosion (AFEX)

The concept of ammonia fiber explosion (AFEX) is similar to steam explosion during which the lignocellulosic materials are exposed to liquid ammonia at high pressure but relatively lower temperatures for a period of time followed by a rapid pressure release (Sun and Cheng, 2002; Talebnia et al., 2010). The ammonia pretreatment is conducted with ammonia loadings around 1-2 kg ammonia/kg dw biomass at moderate temperatures of about 60–100°C and high pressure of about 17–20 atmospheres for a short period of about several minutes (Hendriks and Zeeman, 2009; Amarasekara, 2014). After the ammonia holding period, the vent valve is opened rapidly to explosively relieve the pressure. This rapid release in pressure causes evaporation of the ammonia, which is volatile at atmospheric pressure, and a concomitant sudden drop in temperature of the system occurs (Alizadeh et al., 2005). This process decrystallizes the cellulose, hydrolyses hemicellulose, removes and depolymerises lignin, and increases the size and number of micropores in the cell wall, thereby significantly increasing the rate of enzymatic hydrolysis (Shafiei et al., 2015). The lignin distribution in biomass remains reasonably the same during ammonia AFEX pretreatment, but lignin structure is rigorously altered resulting in increased water-holding capacity and digestibility.

AFEX pretreatment is introduced as one of the leading pretreatment methods for commercialization (Wyman et al., 2005). Several types of reactors have been used in AFEX pretreatment, such as conventional batch reactors, plug-flow reactor (PF-AFEX), packed-bed reactor (PB-AFEX), fluidized gaseous reactors (FG-AFEX) and extractive reactors (E-AFEX) (Shafiei et al., 2015). However, regardless of the reactor type, ammonolysis (amide-forming) and hydrolysis (acid-forming) reactions change the structures of lignin and carbohydrates (Chundawat et al., 2010). The major parameters influencing AFEX process are ammonia loading, temperature, blowdown pressure, moisture content of biomass, and residence time (Holtzapple et al., 1991; Teymouri et al., 2004).

AFEX pretreatment is well suited for various herbaceous crops and agricultural residues with lower amounts of lignin, but it is less efficient for softwood and hardwood. The digestibility of lignocellulosic materials including alfalfa, wheat straw, wheat chaff, barley straw, corn stover, rice straw, municipal solid waste, grasses can be significantly improve by AFEX (Sun and Cheng, 2002). Various agricultural residues after AFEX treatment under optimum conditions could achieve near theoretical sugar yields (Alizadeh et al., 2005; Teymouri et al., 2004). A six-fold increased enzymatic hydrolysis yield and a 2.5-fold ethanol yield after pretreatment by AFEX was reported by Alizadeh et al., (2005). Small particle size is not required for efficacy (Mosier et al., 2005; Sun and Cheng, 2002) allowing a highest solid loading (60–90%) (Wyman et al., 2005) for AFEX pretreatment with anhydrous ammonia among leading pretreatment methods. Moreover, the ammonia pretreatment does not produce inhibitors for the downstream biological processes, so water wash is not necessary (Sun and Cheng, 2002). Both are benefit for the operation of AFEX process and reducing the operational cost. Ammonia is less corrosive compared to acid-catalyzed pretreatments and thus the equipment costs are lower compared with equipment for dilute acid pretreatment. However, the ammonia recovery after the pretreatment is necessary to reduce the cost and protect the environment. The major fraction of ammonia is vaporizes after depressurizing of the materials. Therefore, theoretically it can be recovered and reused again by more than 98% (Chundawat et al., 2013). The recovery of ammonia is possible through compression or cooling and quenching processes (Bals et al., 2011). The economies of these two processes were studied in full biorefinery models and using cooling system was found to be less expensive (Shafiei et al., 2015).

However, ammonia is a toxic, flammable, and volatile substance. Operation of AFEX pretreatment requires technical consideration for a hazardous process, which increases the operating as well as capital costs.

3.2.3. Supercritical Carbon Dioxide Pretreatment

Carbon dioxide behaves as a supercritical fluid at critical temperature (31□C) and critical pressure (7.39 MPa.). CO_2 is cheap, nontoxic, and inflammable. Supercritical CO_2 (SC-CO_2) is becoming an important commercial and industrial solvent due to its ability in solvent extraction and its low toxicity and environmental impact. The properties of SC-CO_2 are widely known and its applications in the pharmaceutical industry are also broad; its high power diffusion and low viscosity represent advantages over conventional solvents. SC-CO_2, as a green solvent suitable for a mobile lignocellulosic biomass processor, has the ability to penetrate the crystalline structure of lignocellulosic biomass overcoming the mass transfer limitations encountered in other pretreatments. Additionally, supercritical fluids show tunable properties such as partition coefficients and solubility (Amarasekara, 2014).

In the supercritical carbon dioxide pretreatment process, SC-CO_2 is delivered to biomass placed in a high pressure container at a pressure of 1000–4000 psi (Zheng et al., 1995). Then the temperature of the pretreatment vessel is increased to about 200□C and held for a short period of time, which will allow CO_2 to penetrate the biomass at high pressure; it was hypothesized that CO_2 would form carbonic acid and increase the hydrolysis rate. Then the pressure is released resulting in the explosive disruption of the lignocellulosic biomass structure, which will increase the accessible surface area for the cellulases in the hydrolysis step (Zheng et al., 1995; Amarasekara, 2014). Due to the release of carbon dioxide at high pressure, lignocelluloses are disturbed, which increases the surface area for further hydrolysis. It is usually known as carbon dioxide explosion pretreatment. Glucose release was observed to increase with increasing pressure and temperature of the carbon dioxide was applied in supercritical carbon dioxide explosion. Explosive steam pretreatment with high-pressure carbon dioxide gave the best overall sugar yields and it increased the glucose yield of bagasse with 50– 70% (Zheng et al., 1998), 14% for yellow pine and 70% for aspen (Kim and Hong, 2001).

Supercritical CO_2 has been tested on woody biomass as well as non-wood soft biomass forms such as wheat straw (Alinia et al., 2010), rice straw (Gao et al., 2010) and corn stover (Lv et al., 2013). The particular behavior of SC-CO2 in the structure of biomass is not well known; however it is known that it works more easily in rigid and wet biomass than in more flexible lignocellulosic materials. Kim and Hong (2001) studied the use of SC-CO_2 pretreatment of hardwood aspen and southern yellow pine with varying moisture contents at various pretreatment conditions followed by enzymatic hydrolysis. They reported significant enhancements in sugar yields in SC-CO_2 pretreated samples when compared to thermal pretreatments. Lv et al., (2013) used SC-CO_2 with water-ethanol as co-solvent to pretreat corn stover to enhance its enzymatic hydrolysis. The results showed that the surface morphology and microscopic structure of pretreated corn stover were greatly changed. Alinia et al., (2010) studied the effect of pretreatment of dry and wet wheat straw by SC-CO_2 alone and by a combination of CO_2 and steam under different operating conditions such as temperature and residence time in the pressure vessels (Alinia et al., 2010). Luterbacher and coworkers have investigated the high pressure (200 bar) CO_2-H_2O pretreatment for various biomass forms such as corn stover, switchgrass, big bluestem, and mixed perennial grasses over a wide range of temperatures (150–250 C) and residence times of 20 seconds to 60 minutes (Alinia et al., 2010). During these studies they found that under these operating conditions, a biphasic mixture of H_2O-rich liquid phase and CO_2- rich supercritical phase coexists and this greatly improves the pretreatment process. Furthermore, they reported that such biphasic pretreatment produced glucose yields of 73% for wood, 81% for switchgrass and 85% for corn stover (Luterbacher et al., 2010).

The use of SC-CO_2 has received attention for biomass pretreatment due to the mild temperatures that can be employed, hence resulting in lesser production of undesirable, degradation, compounds, and because the pretreated material presents no solvent residues, since CO2 is gaseous at environmental conditions. Moreover, there are a number of attractive features in SC-CO_2 pretreatment, which include the low cost of carbon dioxide as a pretreatment solvent, high solid loading, and the use of low temperatures. Nevertheless, the high cost of equipment that can withstand high-pressure conditions of SC-CO_2 pretreatment is a strong limitation to the large-scale application of this process. In addition, the effects on biomass carbohydrate components have yet to be elucidated. Even though a number of investigations of SC-CO_2 with a fairly high degree of success have been reported for pretreatments of various hardwoods and soft biomass forms, the whole process has not proven to be economically viable with the high pressures involved being a deterrent. Improvements are still needed to implement the process on a large scale (Amarasekara, 2014).

3.2.4. Liquid Hot Water (LHW)

Liquid hot water pretreatment (LHW) is also suggested as one of the leading pretreatment methods (Wyman et al., 2005), where pressure is applied to maintain water in the liquid state instead of steam at elevated temperature. Temperature in the range of 170–230°C and P > 5 MPa are commonly used (Sánchez and Cardona, 2008; Talebnia et al., 2010). Other names such as hydrothermolysis, aqueous or steam/aqueous fractionation, uncatalyzed solvolysis, and aquasolv were also used for this type of pretreatment (Shafiei et al., 2015). Unlike in the case of steam explosion applicated in the treatment of non-wood forms of biomass, hot water pretreatment is more suitable for soft biomass forms.

This pretreatment is performed either in batch mode of operation or percolation reactors. There are three types of liquid hot water reactor configurations introduced in the literature (Amarasekara, 2014): (1) flow-through, (2) counter-current, and (3) co-current. Yang and Wyman (2004) discovered that flow through systems removed more hemicellulose and lignin from corn stover than batch systems did, at the same severity factors. During the LHW process, the superheated liquid water auto-ionizes into hydronium ions, which act as a promoter for cleavage of ester bonds of acetyl side chains of hemicelluloses resulting in the formation of acetic acid; the in situ generated acid then autocatalyzes hydrolysis of the hemicellulose and alteration of lignin structure, leading to increased enzyme accessibility to the cellulose fibers (Jiang et al., 2015). The typical pretreatment temperature for percolation-type pretreatment is 190–230°C where the pressure is as high as 20–24 atm. Low solid loadings in range of 2–4% and moderate reaction times of 12–24 min are used in percolation reactors. Products of LHW pretreatment in percolation reactors are a liquid with dilute soluble oligomers and monomers of hemicelluloses and lignin together with a more digestible cellulose riched solid (Wyman et al., 2005; Shafiei et al., 2015). Generally, hot compressed liquid water comes in contact with biomass for up to 15 min, and about 40–60% of the total biomass dissolves in the process, with 35–60% of the lignin, 4–22% of the cellulose, and most of the hemicellulose hydrolyzed to pentosans.

The effectiveness of LHW pretreatment on cellulose digestibility is strongly related to the severity of the conditions. The process severity for LHW pretreatment is defined as (Overend and Chornet, 1987):

$$R_0 = t^* \exp[(T\text{-}100)/14.75] \tag{3.2}$$

where t is reaction time (minutes) and T is the hydrolysis temperature (degrees Celsius). This severity factor is used for both batch and flow through processes. Application of higher process severity increases the solubilization of xylo- and glucooligomers until reaching a maximum. Water at high temperatures is believed to act as an acid, and together with acetyl groups within hemicelluloses are thought to catalyze extensive hydrolysis of hemicellulose to its component monosaccharides and smaller oligosaccharides, primarily xylose. The result of severe pretreatment conditions of LHW is also an accumulation of organic acids such as levulinic acid, acetic acid, and formic acid, which subsequently results in an acidic environment. The acidity in the media can cause degradation of monomeric sugars present in the liquid fraction to compounds such as 5-hydroxymethylfurfural (HMF), and furfural, which are inhibit to the following fermentation step (Amarasekara, 2014). A small amount of mineral acid can be added making the water more acidic, and that would make the process similar to dilute acid pretreatment.

During LHW pretreatment, the pH of the water is affected by temperature and degraded acids, therefore a base like sodium hydroxide can be added to maintain the pH above 4 and below 7 to minimize the formation of monosaccharides. To avoid the formation of inhibitors, maintaining the pH between 4 and 7 is very important to minimize the formation of monosaccharides; and therefore also the formation of degradation products that can further catalyze hydrolysis of the cellulosic material during pretreatment (Hendriks and Zeeman, 2009). If catalytic degradation of sugars occurs, it results in a series of reactions that are difficult to control and result in undesirable side products.

Liquid hot water treatment and steam treatments are similar techniques, however they act differently on the biomass and more efficient than steaming (Laser et al., 2002). Similar to steam explosion, the main advantages of LHW pretreatment are no cost for solvent and neutralization, therefore, this is an environmentally friendly technique and the low cost of the solvent is also an advantage for large-scale application. Furthermore, the equipment cost is reduced since corrosion-resistant materials are not required. Another major advantage in the LHW method is the operation at lower temperatures compared to steam explosion, minimizing the formation of degradation products and the production of inhibitors (Wyman et al., 2005; Yang and Wyman 2004; Mosier et al., 2005).

Then there are disadvantages also in the LHW pretreatment; LHW pretreatment is not efficient for woody biomass (Martin et al., 2011); the amount of solubilized product is higher, while the concentration of these products is lower compared to steam explosion or steam pretreatment (Amarasekara, 2014). This is probably caused by the higher water input in LHW pretreatment compared to steam pretreatment. Therefore, down-stream processing is also more energy demanding since large volumes of water are involved (Amarasekara, 2014). However, on the other hand, due to these lower concentrations the risk on degradation products like furfural and the condensation and precipitation of lignin compounds is reduced.

3.2.5. Ionic liquids

Ionic liquid (IL) pretreatment is a relatively new physicochemical pretreatment technique available for cellulosic ethanol production from biomass (Amarasekara, 2014). They have received considerable attention because they are "designer fluids" that are thermally stable, non-volatile, and capable of dissolving polar and non-polar organic, inorganic, and polymeric com- pounds under mild conditions (Olivier-Bourbigou et al., 2010). Ionic liquids are used as biomass solvents in the pretreatment process and they are green solvents, since they contain no toxics or explosives. They are generally salts with low melting points, existing as liquids below 100°C, and contain large organic cations and small inorganic anions (Rajendran and Taherzaden, 2014). These compounds are well known for their physical properties such as high polarities, high thermal stabilities, and negligible vapor pressure. Cations and anions play different significant roles during pretreatment. Cations interact with lignin by hydrogen bonding and π–π interactions, while anions act as hydrogen bond receptors, interacting with the hydroxyl group of the cellulose, thus disrupting its three-dimensional network (Rajendran and Taherzaden, 2014). The non-hydrated anion like Cl^- is believed to be very effective in breaking the extensive hydrogen-bonding network present, thus bringing a much quicker dissolution and dissolving a higher concentration of cellulose than the traditional solvent systems (Amarasekara, 2014). Different ILs and their specific targets for pretreatment action are summarized by Rajendran and Taherzaden (2014). The most widely used ILs for biomass pretreatment-fraction- ations are 1-allyl-3-methylimidazonium chloride (AMIMCl), 1-^{n}butyl-3-methylimidazonium chloride (BMIMCl), and 1-^{n}butyl- 3-methylimidazonium acetate (BMIMOAc) (Amarasekara, 2014).

Recently, it was demonstrated that ILs can dissolve cellulose, and that cellulose can be re-precipitated after being dissolved in ILs by the addition of an anti-solvent such as water, ethanol or acetone (Zhu et al., 2006). The regenerated biomass showed lower crystallinity than original samples and facile hydrolysis when exposed to cellulase enzymes. The first case about the use of ionic liquids 1-^{n}butyl-3-methylimidazolium (BMIM) for the dissolution of cellulose was reported by Swatloski   et al., (2002). They did not only slowly dissolve high molecular weight pulp cellulose (DP~1000) in [BMIM]$^+$ ionic liquids with Cl$^-$, Br$^-$ and SCN$^-$ anions, but also they reprecipitated the cellulose by adding an anti-solvent like water or alcohol to cellulose-ionic liquid solution. Several other groups have employed this IL-assisted method of cellulose pretreatment to various lignocellulosic materials (Li et al., 2009; Tan and Lee, 2012). Most importantly, the ILs can be recovered and reused after the pretreatment, and different methods such as evaporation, salting out, pervaporation, ion exchange, and reverse osmosis, can be employed for this purpose (Zhu et al., 2006).

Ionic liquid (IL) pretreatment is effective on a variety of biomass forms and some recent applications are in rice straw, wheat straw, and forest and agricultural residues (Amarasekara, 2014). However, using ILs for pretreatment is still expensive that provides a hurdle for its industrialization (Zhu et al., 2006; Mora-Pale et al., 2011). To overcome the economical, cytotoxic, environmental, and energy concerns of conventional imidazolium ILs, new-generation ILs containing ions derived from naturally occurring bases (e.g., choline) and acids (e.g., amino acids and carboxylic acids) have emerged as "completely bio-derived ILs." These ILs contain cholinium cations combined with amino-acid-based anions ([Ch][AA] ILs) or carboxylic-acid-based anions ([Ch][CA] ILs) (Ninomiya et al., 2013). In order to enhance enzymatic saccharification, the ILs method have been also combined use with other pretreatment methods like ultrasound irradiation for the pretreatment of lignocellulosic material (Ninomiya et al., 2013). Therefore, further developments in efficient recovery methods, newgeneration of bio-derived ILs and its combination with other pretreatment methods are vital for any large-scale application of ionic liquid-based pretreatment methods.

3.3. CHEMICAL PRETREATMENT

Chemical pretreatment for lignocellulosic biommass employ different chemicals such as acids, alkalis, oxidizing agents and organosolv (Figure 3.1). Chemical transformations induced by pretreatment reagent(s) are the primary cause of improvement in the accessibility to cellulose and hemicellulose. Among these methods, dilute acid pretreatment using H_2SO_4 is the most-widely used method. Alkaline pretreatment, ozonolysis, peroxide and wet oxidation pretreatments are more effective in lignin removal whereas dilute acid pretreatment is more efficient in hemicellulose solubilization (Talebnia et al., 2010). Chemical transformations are also involved in many physicochemical pretreatments described in Section 3.3 as well.

3.3.1. Acid Pretreatment

The acid pretreatment usually consists of the addition of concentrated or diluted acid solution to the powdered or chopped lignocellulosic biomass, followed by constant mixing at elevated temperatures in the range of 30–220°C (Rajendran and Taherzaden, 2014) and held for a certain time ranging from seconds to minutes at pressure of 3–15 atm (Wyman et al., 2005). Different solid to liquid ratios were studied in batch or continuous operations. Typical solid loading is 10–40%, and solid loading over 50% is not pumpable. (Wyman et al., 2005; Shafiei et al., 2015). The main reaction that occurs during acid pretreatment is the hydrolysis of hemicellulose, especially xylan, making the cellulose more easily accessible for the enzymes, as glucomannan is relatively acid stable. The most commonly used acid is dilute sulfuric acid, but other acids have also been studied such as hydrochloric acid, phosphoric acid, nitric acid (Amarasekara, 2014). Apart from these acids, fumaric and maleic acids have been tested as well. Two modes of acid pretreatment can be performed, e.g., low temperature with high acid concentration and high temperature with dilute acid (Shafiei et al., 2015).

Concentrated-acid (30–70%) hydrolysis of biomass results in high yield of sugars at ambient temperatures, which offers advantage of not using any enzymes for saccharification (Talebnia et al., 2010). Sun et al., (2011) used very high concentrations of sulfuric acid in a continuous, three-step hydrolysis of bamboo biomass, recovering up to 81.6 of the sugars and 90.5% of the acid used. Delgenes et al., (1990) treated wheat straw with 72% (w/v) H_2SO_4 for 30 min at 30 °C and obtained 11.1 g monomeric sugars in total from 18.8 g dry raw material accounting for 59% of maxi- mal theoretical value. However, concentrated acid has drawbacks including high acid and energy consumption, equipment corrosion and longer reaction time (Sun and Cheng, 2002; Taherzadeh and Karimi, 2007, 2008). The concentrated acid must be recovered after hydrolysis to make the process economically feasible (Sun and Cheng, 2002). In addition, strong acid pretreatment is for the ethanol production not attractive, because there is a risk on the formation of inhibiting compounds, which may partially or completely hamper enzymatic hydrolysis and/or fermentation processes. The inhibitor compounds include furfural, 5-hydroxymethylfurfural (HMF), and organic acids such as uronic, formic, levulinic, and acetic acid (Karimi et al., 2013). Therefore, detoxification (Palmqvist and Hahn-Hägerdal, 2000a, b) or application of robust microorganisms is essential for obtaining a high-yield ethanol production (Taherzadeh and Karimi, 2007).

Pretreatment using dilute sulfuric acid at a very low concentration of acid (e.g., 0.1–1.0%) (Rajendran and Taherzaden, 2014) has been considered as one of the most cost-effective methods. The mixture of biomass and dilute acid solution is usually controlled at a higher temperature by means of conventional heating or microwave-assisted heating. High temperature is favorable to attain acceptable rates of cellulose conversion to glucose. Despite low acid concentration and short reaction time, application of high temperatures in dilute-acid hydrolysis accelerates the rate of hemicellulose sugar decomposition and increases equipment corrosion (Taherzadeh and Karimi, 2007; Talebnia et al., 2010). Some equipment such as plug-flow reactors and percolation reactors have been successfully evaluated in laboratory scale or pilot scale. Similar to severity factor used for liquid hot water pretreatment, a modified severity factor is used for dilute acid pretreatment (Yang and Wyman 2004; Shafiei et al., 2015):

$$M0 = t * An^\square * exp[(T-100)/14.75] \qquad (3.3)$$

where A is concentration of acid (wt%), and n is a constant.

Acid pretreatment can provide the efficient hydrolysis of hemicelluloses, partial solubilization of lignin, as well as slight hydrolysis of cellulose by varying the severity of the pretreatment. At elevated temperatures and acidic conditions, hydrogen bonds, which are the major force arranged the cellulose chains together in a crystalline structure, are loosened and acid molecules penetrate inside the cellulose structure. Acid solubilizes a part of acid-soluble lignin which accumulates in the liquid fraction (Xiang et al., 2003) and also cleaves glycosidic bonds in the hemicelluloses as well as glucuronosyl linkages in polyuronides (Jacobsen and Wyman, 2000). Although the pretreatment process is advantageous compared to many other pretreatments, most of the problems arise after the pretreatment process. The main drawbacks of this method are considerable losses of the carbohydrates, operating with low solid to liquid ratios and production of a dilute sugar solution, and formation of many inhibiting by-products and pH neutralization requirement for downstream processes (Sun and Cheng, 2002; Talebnia et al., 2010; Shafiei et al., 2015). Therefore, extensive energy is required for the product recovery and additional processes are essential for detoxification and neutralization of the products, which increase the costs of ethanol production process.

According to Hendriks and Zeeman (2009), acid pretreatment is more attractive for methane production than for ethanol production, because methanogens can handle compounds like furfural and HMF to a certain concentration and with an acclimatization period. For both the ethanol as well as the methane production, the chance on soluble lignin components is a risk, because soluble lignin compounds are often inhibiting for both processes. Methanogens are however capable of adapting to such inhibiting compounds.

3.3.2. Alkaline Pretreatment

Alkaline pretreatment has received a lot of attention lately. Sodium, potassium, calcium, and ammonium hydroxides are commonly used for alkaline pretreatment of lignocelluloses, among which sodium hydroxide has been studied the most (He et al., 2008; Zheng et al., 2009; Talebnia et al., 2010). Alkaline pretreatment processes utilize lower temperatures and pressures than other pretreatment technologies (Mosier et al., 2005). It can be performed at ambient conditions, but longer pretreatment times are generally required rather than at higher temperatures. The alkaline process involves soaking the biomass in alkaline solutions and mixing it at a target temperature for a certain amount of time. A neutralizing step to remove lignin and inhibitors (salts, phenolic acids, furfural, and aldehydes) is required before enzymatic hydrolysis (Amarasekara, 2014). The mechanism of alkaline hydrolysis is believed to be saponification of intermolecular ester bonds crosslinking xylan hemicelluloses and other components, for example, lignin and other hemicellulose (Sun and Cheng, 2002). During alkaline pretreatment the first reactions taking place are solvation and saphonication. This causes a swollen state of the biomass and makes it more accessible for enzymes and bacteria. The porosity of the lignocellulosic materials increases with the removal of the crosslinks (Sun and Cheng, 2002).

This method is well suited for agricultural residues (Sánchez and Cardona, 2008; Sun and Cheng, 2002). Sodium hydroxide has received the most attention as it is less expensive than others, and there are extensive studies on using sodium hydroxide on various biomass forms (Zheng et al, 2009; Xu and Cheng, 2011; Kang et al., 2012). Sodium hydroxide pretreatment is usually carried out by treating biomass with 1–12% aqueous sodium hydroxide, typically at 50–120°C; the exposure time may vary from minutes to a few days. He et al., (2008) investigated the biogas yield of rice straw during anaerobic digestion through solid-state sodium hydroxide pretreatment. The biogas yield of 6% NaOH-treated rice straw was increased by 27.3-64.5%. The use of NaOH in the pretreatment caused saponification of intermolecular ester bonds crosslinking lignin (Figure 3.2) and part of the hemicellulose, resulting in structural alteration of lignin. In addition to this, partial decrystallization of cellulose and partial solvation of hemicellulose can occur during the alkaline pretreatment process. The possible reactions are shown in Figure 3.2. Zheng et al., (2009) advanced a wet state sodium hydroxide method to pretreat corn stover for enhancing biogas production. The NaOH dose of 2% (NaOH weight to dry biomass) and the loading of 65 g/L were found to be optimal in terms of 72.9% more total biogas production, 73.4% more methane yield, and 34.6% shorter technical digestion time, as compared to the untreated one. As different dosages of NaOH were used, the main compositions of corn stover changed obviously (Zheng et al., 2009). Although NaOH has been shown to effectively enhance cellulose digestibility, it has several disadvantages, such as cost, safety concerns, and difficult to recover.

Lime has received much more attention since it is inexpensive (about 6% cost of sodium hydroxide), has improved handling, and can be recovered easily by using carbonated wash water. Lime is a week alkaline agent and has partial solubility in water (< 0.2%) usually used in alkali pretreatment. Solubility of calcium hydroxide decreases as the temperature is increased. Lime pretreatment was used to improve in vitro digestibility of crop residues used for animal feed. Lime pretreatment solubilizes about 33% of lignin and 26% of xylan. In lime pretreatment, biomass slurry (5–15 g H_2O/g dry biomass) is pretreated by calcium hydroxide (> 0.1 g $Ca(OH)_2$/g dry biomass). Long-term lime pretreatment is performed at mild temperatures (25–60°C) for several days to weeks (Sierra et al., 2010). This type of pretreatment is suggested for treating biomass piles, in which no specific pretreatment reactor is required. Percolation of air as an inexpensive oxidizing agent is possible in long-term lime pretreatment (Wyman et al., 2005).

Lime pretreatment are efficient for herbaceous materials with low-lignin content (Chang et al., 1997; Gandi et al., 1997; Kaar and Holtzapple, 2000; Wyman et al., 2005), but not for biomass with higher lignin content such as poplar (hardwood) (Chang et al., 2001b). One advantage of lime pretreatment is that the cost of materials required to pretreat a given quantity of biomass is the lowest and higher safety compared with other alkaline reagents (Chang et al., 1997). The carbon dioxide required for neutralization of lime is produced in fermentation and no additional cost is required for neutralizing agent. The produced calcium carbonate is not soluble in water and its separation is easier than other salts. Furthermore, lime can be recovered as insoluble calcium carbonate $CaCO_3$ by precipitation with CO_2 after solid-liquid separation. The calcium carbonate can then be converted to quicklime (CaO) using well established lime kiln technology. As a further advancement in this direction, Park and coworkers developed a novel lime-pretreatment process (CaCCO process) that did not require a solid- liquid-separation step (Park et al., 2010).

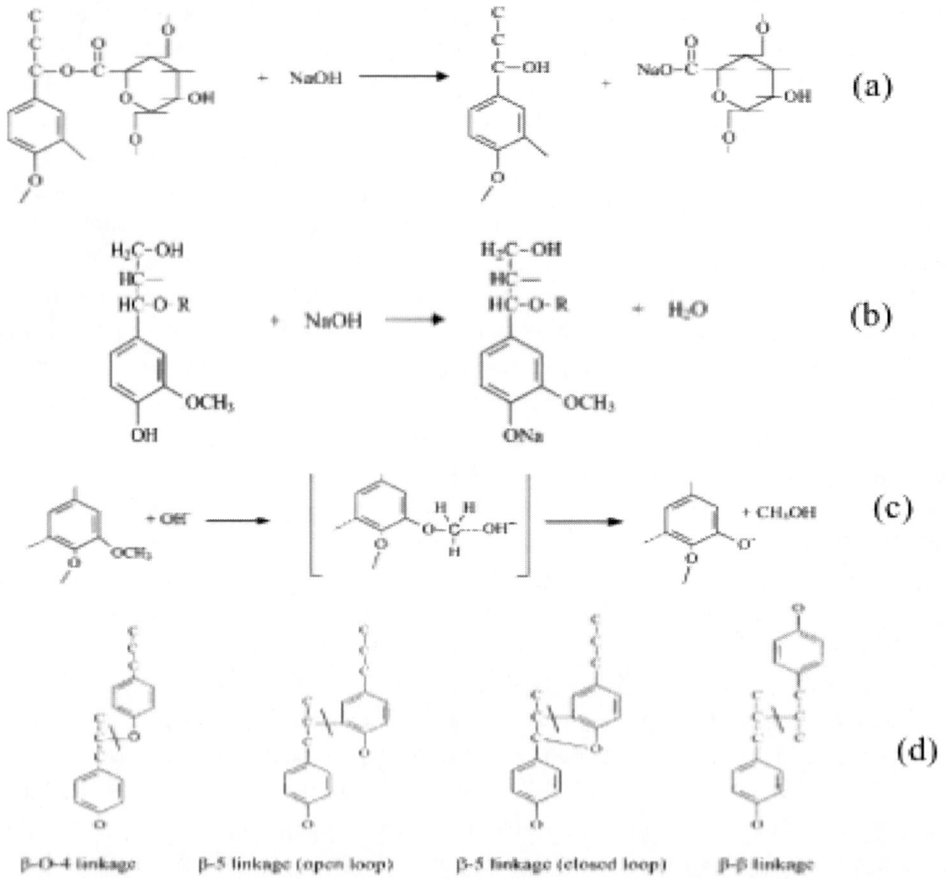

Figure 3.2. The reactions during NaOH pretreatment: (a) Scheme of lignin carbohydrate complexes reacting with NaOH; (b) Scheme of lignin reacting with NaOH; (c) Scheme of methoxyl separating from the lignin structure; (d) disrupt of lignin structure linkages.

One of the disadvantages of lime pretreatment is slower reactions compared with other alkaline pretreatments. Addition of oxidizing agents may lead to non-selective degradation reactions. Losses of carbohydrates as well as production of inhibitor aromatic compounds may occur as well (Hendriks and Zeeman, 2009; Shafiei et al, 2015). A number of research groups have studied combinations of alkaline pretreatment with other pretreatment methods such as wet oxidation, steam explosion, ammonia fiber explosion as described in this chapter, which may overcome these problems.

3.3.3. Oxidative Pretreatment

An oxidative pretreatment consists of the addition of an oxidizing agents such as ozone (Alvira et al., 2010), hydrogen peroxide, and oxygen, to the biomass. The objective is to remove the hemicellulose and lignin to increase the accessibility of the cellulose. This pretreatment is usually combined with other chemical or hydrothermal treatments.

(1) Hydrogen Peroxide Oxidation

In the oxidation by hydrogen peroxide, aromatic and olefinic structures are destroyed by a nucleophilic attack of hydrogen peroxide anions (Zhao et al., 2012; Rajendran and Taherzaden, 2014). During hydrogen peroxide oxidative pretreatment several reactions may take place, like electrophilic substitution, displacement of side chains, cleavage of alkyl aryl ether linkages or the oxidative cleavage of aromatic nuclei (Hon and Shiraishi, 2001). Hydrogen peroxide delignification of agricultural residues showed to be strongly depended on pH, with an optimum pH of 11.5–11.6. Gould (1984) demonstrated the use of H_2O_2 for delignification with a maximum at pH 11.5. Some researchers reported that at pH lower than 10, delignification was negligible and at a pH 12.5 and higher, hydrogen peroxide had no real effect on the enzymatic digestibility (Gould, 1984; Talebnia et al., 2010). According to Gould (1984) the hydrogen peroxide concentration should be at least 1% and the weight ratio between H_2O_2 and biomass should be 0.25 for a good delignification. The delignification is probably caused by the hydroxyl ion (HO), which is a degradation product of hydrogen peroxide with a maximum at pH 11.5–11.6. About half of the lignin was solubilized in this way (temperature around 25 °C and a duration of 18–24 h). In many cases the used oxidant is not selective and therefore losses of hemicellulose and cellulose can occur. At the same time, lignin is oxidized and soluble aromatic compounds are formed, so a high risk on the formation of inhibitors exists (Hendriks and Zeeman, 2009).

(2) Wet Oxidation (WO)

Wet oxidation is a chemical pretreatment technique that involves exposure of biomass to water using air or oxygen as oxidizing agents at temperatures above 120 °C, sometimes with the addition of an alkali catalyst. The operating temperature is usually in the range of 125–320°C, with residence time, approximately 30 minutes. Typical oxygen pressure range is 0.5–2 MPa (Amarasekara, 2014). Temperature, reaction time, and the oxygen pressure are considered to be the crucial parameters in this pretreatment method (Schmidt and Thomsen, 1998). The main reactions in WO pretreatment comprise formation of acids and oxidative reactions (Rajendran and Taherzaden, 2014).

Wet oxidation attacks all fractions of the lignocelluloses. Hemicelluloses are decomposed into monomeric sugars. Lignin is both oxidized and cleaved, and cellulose is degraded to a certain extent. (Rajendran and Taherzaden, 2014). This method is suited to materials with low lignin content, since the yield has been shown to decrease with increased lignin content, and since a large fraction of the lignin is oxidized and solubilized to carbon dioxide, water, and carboxylic acids (Banerjee et al., 2009). The amount of lignin removed in the WO process ranges from 50% to 70% depending on conditions used and type of biomass pretreated. As with many other delignification methods, the lignin cannot be used as a solid fuel, which considerably reduces the income from by-products in large-scale production. The WO technique is very effective in removing dense wax coating containing silica and protein in biomass forms such as straw, reed, and other cereal crop residues (Schmidt et al., 2002; Qiang, and Thomsen, 2012; Arvaniti et al., 2012). Banerjee et al., (2009) obtained the best WO pretreatment conditions for rice husk where they applied a reaction temperature of 185°C and air pressure of 0.5 MPa for 15 min. 66.97% of the cellulose in the solid fraction were maintained, while 89% lignin was removal. On the other hand, 69.77% of the hemicellulose was solubilized through the WO treatment. However, the amount of byproducts formed was almost always higher for pretreatment by WO than by steam explosion (Martín et al., 2008). Byproducts obtained included succinic acid, glycolic acid, formic acid, acetic acid, phenolic compounds, and furfural, which would have negative effects on further downstream processing due to inhibition (Amarasekara, 2014).

The benefits of combining wet oxidation with other pretreatment methods to further increase the yield of sugars after enzymatic hydrolysis were studied (Sorensen et al., 2008; Georgieva et al., 2008). Combination of alkali and WO not only improves the rate of lignin oxidation (and in turn enzymatic hydrolysis) but also prevents formation of furfural and HMF (Talebnia et al., 2010). Bjerre et al., (1996) investigated the alkaline wet oxidation of wheat straw at 10 bar of oxygen pressure and various temperatures. A maximum delignification of about 65% and hemicellulose solubilization of about 50% occurred at 170 °C within 10 min. The left over solid residue showed the highest enzymatic convertibility where 85 (%w/w) of cellulose was converted to glucose. A combination of wet oxidation and steam explosion is also known as wet explosion. In this technique the biomass not only undergoes the chemical reaction described above, but also undergoes physical rupture due to steam explosion (Sorensen et al., 2008). The main advantages in combining wet oxidation with steam explosion is the ability of the combined process to handle larger pieces of biomass and to operate at higher biomass loadings per given volume of pretreatment solvent medium (Georgieva et al., 2008; Amarasekara, 2014).

Energy requirements for wet oxidation pretreatments are relatively low since the only thermal energy required for these pretreatments is the difference in enthalpy between the incoming and outgoing streams. However, the capital cost for wet oxidation pretreatments is higher than other pretreatment techniques, and the operating costs are mainly the power to produce compress air (Amarasekara, 2014).

(3) Ozonolysis

Ozone is a powerful oxidant that is well known in the water treatment industry and in the paper industry for pulp bleaching. Ozonolysis pretreatment is an effective method and is specifically used for degradation of lignin and partial degradation of hemicelluloses. Many lignocellulosic materials such as loblolly pine, sweetgum, cotton stalks, wheat straw, Miscanthus, and poplar sawdust can be pretreated by ozone (Vidal and Molinier, 1988; García-Cubero et al., 2009; Sannigrahi et al., 2012).

Ozone is highly reactive towards compounds with conjugated double bonds and functional groups with high electron densities. Therefore, the moiety most likely to be oxidized in ozonization of lignocellulosic biomass materials is lignin due to its high content of carbon-carbon double bounds (Amarasekara, 2014). In this process, ozone is essentially limited to solubilize lignin and a small fraction of hemicellulose, but cellulose was hardly affected (Talebnia et al., 2010). Ozone attacks lignin releasing soluble compounds of smaller molecular weight, mainly organic acids such as formic and acetic acid, which can result in a drop in pH from 6.5 to about 2. Vidal and Molinier reported that the rate of enzymatic hydrolysis of wheat straw increased by a factor of 5 following 60% removal of the lignin from wheat straw by ozone pretreatment (Vidal and Molinier, 1988).

Ozonolysis is mostly carried out at room temperature and pressure (Vidal and Molinier, 1988). Important parameters to consider in ozonolysis pretreatment are moisture content of sample, particle size, and concentration of ozone(Rajendran and Taherzaden, 2014), among which, moisture content found to be the most significant variable and a reaction controlling parameter for values below 30%. Prominent advantages of ozonolysis are that no inhibitory compounds are formed, and also no acids, alkali, or toxic compounds, making it a very attractive choice of pretreatment (Vidal and Molinier, 1988; Sun and Cheng, 2002). The main drawback of this process is a large amount of ozone utilization that makes the process expensive, which is a challenge to its feasibility (Sun and Cheng, 2002; Talebnia et al., 2010).

3.3.4. Organosolv Pretreatment

Organosolv pretreatment (OP) process is a pretreatment means for lignocellulosic feedstock, in which biomass is treated with organic solvents and water with or without added acid/base catalyst usually at 100–250°C for a short time. The exact temperature depends on the type of organic solvents used. A number of solvents i.e., methanol, ethanol, acetone, ethylene glycol, triethylene glycol, glycerol, and tetrahydro-furfuryl alcohol with or without the addition of a catalyst agent have been used as the organic solvent in this process. The catalysts used include organic or inorganic acids (hydrochloric and sulfuric acids), bases (sodium hydroxide, ammonia and lime). Low boiling point alcohols, such as methanol and ethanol, appear to be the most suitable organic liquids for use in organosolv processes, due to their low cost and facile recovery, and ethanol is safer and preferred over methanol due to toxicity (Amarasekara, 2014). Besides, acetone is also a widely used organic solvent for pretreatment of lignocelluloses (Rajendran and Taherzaden, 2014). Many forms of woody and softer non-wood biomass materials have been tested in the organosolv process. A major function of organosolvs is hydrolysis the internal lignin bonds, as well as the ether and 4-O-methylglucuronic acids ester bonds between lignin and hemicellulose; additionally, glycosidic bonds in hemicellulose are also partially hydrolyzed during the organosolv pretreatment depending on process conditions (Zhao et al., 2009). The complete removal of lignin and hemicelluloses results in a highly accessible surface area, and large pore volumes (Zhao et al., 2012).

The addition of a catalyst for example dilute sulfuric acid, oxalic, salicylic, and acetylsalicylic acid, can enhance the selectivity of the solvent with respect to lignin or to decrease the operating temperature (Taherzadeh and Karimi, 2008; Huijgen et al., 2011). For instance, sodium hydroxide as a catalyst agent during organosolv ethanol pretreatment is known to improve the selectivity with respect to lignin. This may be due to the fact that ethanol can reduce the surface tension of the pulping liquor, favoring the alkali penetration into the material, consequently removing the lignin. In addition, organosolv pretreatment can be combined with other pretreatment techniques to obtain a clean and effective biomass fractionation process or multi-stage pretreatment processes for more recalcitrant biomass as means for improving pretreatment yield (Amarasekara, 2014).

The organosolv pretreatment method is very effective for the pretreatment of high-lignin lignocellulosic materials such as soft woods. The output of organosolv pretreatment consists of dry lignin, aqueous hemicellulose, and relatively pure cellulose. The three separated fractions can be utilized or converted to feedstock for the chemical industry (Amarasekara, 2014). The hemicellulose sugars recovered from the water-soluble stream can be concentrated for fermentation using special organisms to convert the five-carbon aldose to ethanol or other products. In particular, lignin separated in the process is sulfur free with high purity, low molecular weight and abundance of reactive groups. It can be used as a fuel to power cellulosic ethanol plants, or further purified to obtain high quality lignin, or as feedstock for the production of phenolic resins/adhesives, antioxidant, bio-based polymer composites, and even hydrocarbon products for blending with gasoline (Tian et al., 2013; Amarasekara, 2014).

As described above, the flexibility and good generality combined with the ease of recovering the lignin and hemicellulose sugar streams make organosolv pretreatment one of the most efficient pretreatment techniques. However, organosolv pretreatment still possesses several disadvantages.

As volatile organic liquids are used at high temperature and pressure it is essential to use high-quality containment vessels, since pretreatment-digester leaks can cause fire, explosion hazards, environmental and health and safety concerns. The cost of organosolvs and catalyst makes OP process more costly than other pretreatment processes, therefore, recovery and recycling of the solvent is necessary to reduce costs. As the solvent itself and the furfural and 5-hydroxymethylfurural generated by acid-catalyzed degradation of monosaccharides might inhibit the enzymatic hydrolysis, removal of the solvent after the pretreatment is compulsory (Sun and Cheng, 2002; Amarasekara, 2014).

3.4. BIOLOGICAL PRETREATMENT

Biological pretreatment is considered to be a "green" technology as it is performed under ambient conditions without chemical addition. The main benefits include cost effective, low energy requirements, environment friendly, little or no waste stream output, and low formation of toxic materials such as furfural, hydroxymethylfurfural, etc. It has the potential to be applied to on-farm wet storage for cost-effective biofuels production from lignocellulosic biomass. Biological approach has been demonstrated using direct microorganism as well as using enzymes extracted from microbes. This section presents a brief review of fungi and enzymes involved in biological pretreatment of lignocellulosic biomass.

3.4.1. Fungi

The most common type of microorganisms used in this pretreatment is fungi such as white-rot fungi, brown-rot fungi and soft-rot fungi. Both of brown-rot and soft-rot fungi principally degrade the plant polysaccharides with minimal lignin degradation, while white-rot fungi are capable of complete mineralization of both the lignin and the polysaccharide components (Rajendran and Taherzaden, 2014).

The biodegradation of lignocelluloses by fungi is associated with their mycelial growth. This habit allows fungi to transfer rare nutrients such as nitrogen and iron to a nutrient-poor lignocellulosic substrate that constitutes carbon source. Fungal degradation occurs extracellularly by association with outer cell envelope layer. There are two types of extracellular enzymatic systems: (a) hydrolytic system and (b) ligninolytic system. In hydrolytic system, hydrolases are produced for degrading polysaccharides, while ligninolytic system opens phenyl rings and degrades lignin. *Trichoderma reesei* is one of the widely studied fungi used in commercial production of cellulases and hemicellulases (Sánchez, 2009; Rajendran and Taherzaden, 2014).

Some white-rot fungi, such as *Phanerochaete chrysosporium, Ceriporia lacerata, Cyathus stercolerus, Ceriporiopsis subvermispora, Pycnoporus cinnabarinus, Phlebia subserialis, Echinodontium taxodii,* and *Pleurotus ostreatus* show high efficiency in terms of removing lignin (Yu et al., 2009; Alvira et al., 2010; Xu et al., 2010; Isroi et al., 2011). Most white-rot fungi grow well on hard woods such as birch and aspen, while certain species *Heterobasidion annosum, Phellinus pini,* and *P. radiata* grow well on soft woods such as pine and spruce (Isroi et al., 2011).

These fungi mainly produce ligninolytic enzymes and H_2O_2, when entering the Fenton reaction results in release of free radicals. The radicals attack the lignin walls, and they split open, allowing the enzymes to perform. Oxidative reactions cause depolymerization of lignin. The final products would be carbon dioxide and water (Yu et al., 2009; Ma et al., 2010; Isroi et al., 2011; Zhao et al., 2012).

The growth of fungi on lignocellulosic biomass results in a loss of dry matter. During the fungal growth, all the main components (cellulose, hemicelluloses, and lignin) are consumed in part by the fungus for its growth and metabolic activities. White-rot fungi degrade lignin in biomass with two different mode of degradation, named as selective and non-selective degradation.

In non-selective degradation, all three components (lignin, cellulose, and hemicellulose) were almost degraded equally, whereas in selective decay mostly hemicellulose and lignin were degraded (Isroi et al., 2011). Although it is very difficult to remove lignin alone from the lignocellulose, more than 1,500 species of white-rot fungi are able to decompose lignin with little consumption of cellulose (Tian et al., 2012), such as *Stropharia rugosoannulata, Hapalopilus rutilans, P. ostreatus, C. subvermispora, Lentinula edodes,* and *Pleurotus eryngii.*

They are able to consume lignin faster than non-lignin content of biomass. Therefore, these strains are good delignifier and can be used efficiently in biological pretreatment of lignocellulose (Narayanaswamy et al., 2013).

Biological pretreatment by fungi is environmentally friendly, low energy consumption, modest environmental conditions and no chemical requirement. This makes it economically feasible. However, biological pretreatment still faces some drawbacks negatively affecting its widespread application as a commercial pretreatment method. These include long process time, large space requirement and the need for continuous monitoring of microorganism growth (Haghighi Mood et al., 2013).

An additional challenge is that some cellulose and hemicellulose are consumed during the pretreatment, which reduces the total content of reducing sugars after the pretreatment (Rajendran and Taherzaden, 2014).

3.4.2. Enzymes

Although several basidiomycetes strains were shown to be able to reduce the lignin content of lignocellulosic biomass, the rate of fungal delignification was too low for industrial use (Woiciechowski et al., 2013).

Enzymatic treatments, which take only a few hours and are very selective, are therefore more suitable than fungal treatments. There are four major groups of ligninolytic enzymes produced by the white-rot fungi: lignin peroxidase (LiP; 1, 2-bis (3,4-dimethoxyphenyl) propane-1, 3-diol: hydrogen-peroxide oxidoreductase; EC 1.11.1.14), manganese dependent peroxidase (MnP; Mn(II): hydrogen-peroxide oxidoreductase or manganese peroxidase; EC 1.11.1.13), versatile peroxidase (VP; EC 1.11.1.16), and laccase (benzenediol: oxygen oxidoreductase; EC 1.10.3.2) (Narayanaswamy et al., 2013; Woiciechowski et al., 2013).

Many of the white- and brown-rot fungi produce enzymes which are competent to degrade lignin, such as laccase, manganese peroxidase, and versatile peroxidase. Nonetheless, lignin peroxidase, is not produced by many white-rot fungi (Isroi et al., 2011). Very few fungi are found to produce extracellular LiP. *P. chrysosporium*, *T. versicolor*, *Bjerkhandera* sp., and *T. cervina* are some fungi, which can produce LiPs (Isroi et al., 2011). LiPs are monomeric homo-protein and glycol protein belonging to oxidoreduc- tase family, which specifically act on peroxide as an acceptor (peroxidases). These enzymes have molecular weight of 40 kDa and isoelectric points (pI) ranging from 2.8 to 5.3. The interaction of LiPs with its substrate follows Ping-Pong mechanism. LiPs oxidize non-phenolic and phenolic units of lignin by removing one electron and creating free radicals, which lead to chemically decompose the polymer.

LiP has been shown to oxidize fully methylated lignin, lignin model compounds as well as various polyaromatic hydrocarbons. LiPs cleave selectively $C\alpha$–$C\beta$ bond, aryl $C\alpha$ bond, aromatic ring opening and demethylation in the lignin molecule (Isroi et al., 2011; Narayanaswamy et al., 2013).

Manganese peroxidase also require H_2O_2 as an oxidant in the Mn-dependent catalyzing reaction in which Mn^{2+} is converted to Mn^{3+} by MnP. Mn^{3+} then oxidizes phenolic rings to phenoxyl radicals, which leads to decomposition of compounds. *P. chrysosporium*, *Pleorotus ostreatu*, *Trametes* sp., and several other species belong to *Meruliaeiae*, *Coriolaceae*, and *Polyporaceae* produce MnP (Isroi et al., 2011). Both LiP and MnP belong to the class of peroxidases that oxidize their substrates by two consecutive one-electron oxidation steps with intermediate cation radical formation. Due to its high redox potential, the preferred substrates for LiP are nonphenolic methoxyl-substituted lignin subunits and the oxidation occurs in the presence of H_2O_2 (Wong, 2009; Woiciechowski et al., 2013) whereas MnP acts exclusively as a phenol oxidase on phenolic substrates using Mn^{2+}/ Mn^{3+} as an intermediate redox couple (Tuor et al., 1995). MnPs have an important role in lignin depolymerization, chlorolignin, and demethylation of lignin. Therefore, MnPs have a very essential role in biological pretreatment of lignocellulosic biomass.

Laccase has been found in fungi, bacteria, and plants. The major producers of laccase are of fungi kingdom. Laccase is generally larger than peroxidases as it has a molecular weight of approx- imately 60 kDa and pI 3–6 (Hatakka, 2001). Laccases belonging to blue multicopper protein or oxidase family are able to oxidize a variety of phenolic compounds including polyphenols, methoxy-substituted phenols, diamines and a considerable range of other compounds, with concomitant reduction of molecular oxygen to water (Dwivedi et al., 2011). Optimum pH for better Laccase activity is found to be 3–5 (Heinzkill et al., 1998). They oxidize phenols and phenolic lignin substructures by one-electron abstraction with formation of radicals that can repolymerize or lead to depolymerization (Woiciechowski et al., 2013). It also have been found to oxidizes non-phenolic compounds of lignin with addition of ABTS in the presence of a mediator, such as 2,2'-azinobis (3-ethylbenzthiazoline-6-sulfonate) (ABTS),

1-hydrobenzotriazole (1-HBT), and violuric acid (Wong, 2009; Woiciechowski et al., 2013)). Therefore, the natural mediator should be produced by organisms for the complete oxidation of lignin.

The highest amounts of laccases are produced by white-rot fungi. Fungal laccases are secreted into the medium by the mycelium of filamentous fungi (Couto and Toca-Herrera, 2007). Examples of microorganisms that produce laccase with high activity are Trametes pubescens (740,000 U/L) (Galhaup et al., 2002).

Versatile peroxidases are a group of enzymes found in various *Bjerkandera* species and *Pleurotus* species (Woiciechowski et al., 2013). They are primarily recognized as manganese peroxidases, which exhibit activities on aromatic substrates similar to that of LiP. VP can oxidize both phenolic and non-phenolic compounds of lignin as well as Mn^{2+}(Isroi et al., 2011). These enzymes are not only specific for Mn(II), but also oxidize phenolic and non-phenolic substrates including veratryl alcohol, methoxybenzenes, and lignin model compounds in the absence of manganese (Wong, 2009). The catalytic mechanism is similar to LiP.

Significant amounts of these enzymes have been produced by submerged fermentation and solid-state fermentation. Ligninolytic enzymes principally target at lignin by the oxidative reaction in the presence of mediators. These enzymatic reactions are carried out at 30–45°C with low enzyme loading rate for about 6–26 hours. More research is happening recently in the purification of these enzymes and improving the activity of enzymes by mutagenesis and rDNA technology (Hao et al., 2006; Mtui and Nakamura, 2010; Mtui, 2010; Rajendran and Taherzaden, 2014).

3.5. OVERVIEW OF PRETREATMENT METHODS

Pretreatment plays a significant role in biofuel production from lignocellulosic materials. The objectives are to increase the surface area and porosity of the substrate, reduce the crystallinity of cellulose and disrupt the heterogeneous structure of cellulosic materials. Table 3.3 describes some the most promising methods and corresponding effects discussed in this chapter. The table suggests that increasing the surface area is one of the major approaches of a pretreatment by solubilization of the hemicellulose and/or lignin and/or altering the lignin.

Diverse advantages and drawbacks are associated with each pretreatment method.

A comparison of major pretreatment methods is presented in Table 3.4. It is difficult to define the best pretreatment method as it depends on many factors such as type of lignocellulosic biomass and desired products. Pretreatments have one or more effects on substrates; however, thus far, no single pretreatment method was found to meet all the pretreatment requirements; instead a combination of different methods might be applied. The other challenge of production of fuels, as ethanol, methane and hydrogen from the treated lignocellulosic materials is to make it as economical as possible with the new technologies that have been developed recently.

Table 3.3. Effects of the different pretreatments on the physical/chemical composition or structure of lignocellulose (adapted from Hendriks and Zeeman, 2009)

Pretreatment method	Increase accessible surface area	Decrystallization on cellulose	Solubilization hemicellulose	Solubilization lignin	Inhibitor formation	Alteration lignin structure
Mechanical	+	+				
SE	+		+	-	+	+
LHW	+	ND	+	+/-	-	-
Acid	+		+	-	+	+
Alkaline	+		-	+/-	-	+
Oxidative	+	ND		+/-	-	+
Thermal +acid	+	ND	+	+/-	+	+
Thermal+oxidative	+	ND	-	+/-	-	+
AFEX	+	+	-	+	-	+
CO$_2$ explosion	+		+			
Ionic liquids	+	+	+	+	-	+
Biological	+	-	+	+	-	+

+ = major effect; - = minor effect; ND = not determined.

Table 3.4 A comparison of major pretreatment methods (Adapted from Amarasekara, 2014)

Pretreatment method	Inhibitor formation	Reuse of chemicals	Applicability to different feedstock's	Equipment cost	Success at pilot scale	Advantages	Limitations & disadvantages
Mechanical	Nil	No	Yes	H	Yes	Reduce cellulose crystallinity	High power and energy consumption
Acids	H	Yes	Yes	H	Yes	Hydrolysis of cellulose and hemicellulose, alters lignin structure	Hazardous, toxic and corrosive
Dilute aqueous NaOH	L	Yes	Yes	L	Yes	Removal of lignin and hemicellulose, increases accessible surface area	Long residence time, irrecoverable salts formed

Pretreatment method	Inhibitor formation	Reuse of chemicals	Applicability to different feedstock's	Equipment cost	Success at pilot scale	Advantages	Limitations & disadvantages
Liquid hot water	H	No	-	-	Yes	Removal of hemicellulose making enzymes accessible to cellulose	Long residence time, less lignin removal
Organosolv	H	Yes	Yes	H	Yes	Hydrolyze lignin and hemicellulose	High cost Solvents needs to drained, and reused
Wet oxidation	Nil	No	-	H	-	Removal of lignin, low formation of inhibitors and minimize energy demand	High cost of oxygen and alkaline catalyst
Ozonolysis	L	No	-	H	No	Reduces lignin content, no toxic residues	High cost of large amount of ozone requied
CO_2 explosion	L	No	-	H	-	Hemicellulose removal, cellulose decrystallization, cost-effective No toxic compounds produced	Does not modify lignin High pressure
Steam explosion	H	-	Yes	H	Yes	Hemicellulose removal and alteration in lignin structure	Incomplete destruction of lignin carbohydrate matrix and generation of inhibitors
AFXE	L	Yes	-	H	-	Increases accessible surface area Low formation of inhibitors	Not efficient for biomass with high lignin content High cost of large amounts of ammonia
Ionic liquids	L	Yes	Yes	-	-	Dissolution of cellulose, increased amenability to cellulose	Still in initial stage
Fungi	L	-	Yes	L	-	Degradation of lignin and hemicellulose leaving cellulose intact	Low rate of hydrolysis
Enzymatic	Nil	No	Yes	L	Yes	Higher hydrolysis rate	High cost of enzymes

REFERENCES

Aden, A., Ruth, M., Ibsen, K., Jechura, J., Neeves, K., Sheehan, J., Wallace, B., Montague, L., Slayton, A. & Lukas, J. (2002). Lignocellulosic biomass to ethanol process design and economics utilizing co-current dilute acid prehydrolysis and enzymatic hydrolysis for corn stover. National Renewable Energy Laboratory. US Department of Energy Laboratory.

Aden, A. & Foust, T. (2009). Technoeconomic analysis of the dilute sulfuric acid and enzymatic hydrolysis process for the conversion of corn stover to ethanol. *Cellulose*, *16*(4), 535–545.

Alinia, R., Zabihi, S., Esmaeilzadeh, F. & Kalajahi, J. F. (2010). Pretreatment of wheat straw by supercritical CO_2 and its enzymatic hydrolysis for sugar production. *Biosystems Engineering*, *107*(1), 61–66.

Alizadeh, H., Teymouri, F., Gilbert, T. I. & Dale, B. E. (2005). Pretreatment of switchgrass by ammonia fiber explosion (AFEX). *Applied Biochemistry and Biotechnology - Part A Enzyme Engineering and Biotechnology*, *124*(1–3), 1133–1141.

Amarasekara, A.S. (2014). Pretreatment of Lignocellulosic Biomass. In: *Handbook of Cellulosic Ethanol*. Scrivener Publishing LLC, John Wiley & Sons, Inc, 147-217 (Chap.5).

Avellar, B. K. & Glasser, W. G. (1998). Steam-assisted biomass fractionation. I. Process considerations and economic evaluation. *Biomass Bioenerg.* *14*(3), 205–218.

Alvira, P., Tomás-Pejó, E., Ballesteros, M. & Negro, M. J. (2010). Pretreatment technologies for an efficient bioethanol production process based on enzymatic hydrolysis: a review. *Bioresour. Technol.*, *101*(13), 4851–4861.

Agbor, V., Carere, C., Cicek, N., Sparling, R. & Levin, D. (2014). Biomass pretreatment for consolidated bioprocessing (CBP). In: Waldron K (ed) *Advances in biorefineries*. Woodhead Publishing, Sawston, 234–258 (Chap. 8).

Arvaniti, E., Bjerre, A. B. & Schmidt, J. E. (2012). Wet oxidation pretreatment of rape straw for ethanol production. *Biomass Bioenerg.*, *39*, 94–105.

Azuma, J., Tanaka, F. & Koshijima, T. (1984). Enhancement of enzymatic susceptibility of lignocellulosic wastes by microwave irradiation. *Journal of Fermentation Technology*, *62*, 377–384.

Ballesteros, I., Oliva, J. M., Navarro, A. A., Gonzalez, A., Carrasco, J. & Ballesteros, M. (2000). Effect of chip size on steam explosion pretreatment of softwood. *Appl. Biochem. Biotechnol.*, 84–86, 97–110.

Ballesteros, I., Negro, M. J., Oliva, J. M., Cabanas, A., Manzanares, P. & Ballesteros, M. (2006). Ethanol production from steam-explosion pretreated wheat straw. *Appl. Biochem. Biotechnol.*, *130*, 496–508.

Bals, B., Wedding, C., Balan, V., Sendich, E. & Dale, B. (2011). Evaluating the impact of ammonia fiber expansion (AFEX) pretreatment conditions on the cost of ethanol production. *Bioresour. Technol.*, *102*(2), 1277–1283.

Banerjee, S., Sen, R., Pandey, R. A., Chakrabarti, T., Satpute, D., Giri, B. S. & Mudliar, S. (2009). Evaluation of wet air oxidation as a pretreatment strategy for bioethanol production from rice husk and process opti- mization. *Biomass Bioenerg.*, *33*(12), 1680–1686.

Beltrame, P. L., Carniti, P., Visciglio, A., Focher, B. & Marzetti, A. (1992). Fractionation and bioconversion of steam-exploded wheat straw. *Bioresour. Technol.*, *39*, 165–171.

Bjerre, A. B., Olesen, A. B., Fernqvist, T., Ploger, A. & Schmidt, A. S. (1996). Pretreatment of wheat straw using combined wet oxidation and alkaline hydrolysis resulting in convertible cellulose and hemicellulose. *Biotechnol. Bioeng.*, *49*, 568–577.

Brodeur, G,. Yau, E., Badal, K., Collier, J., Ramachandran, K. B. & Ramakrishnan, S. (2011). Chemical and physicochemical pretreatment of lignocellulosic biomass: a review. *Enzyme Res.*, 2011, *17*.

Brownell, H. H., Yu, E. K. C. & Saddler, J. N. (1986). Steam-explosion pretreatment of wood: effect of chip size, acid, moisture content and pressure drop. *Biotechnol. Bioeng.*, *28*, 792–801.

Cadoche, L. & López, G. D. (1989). Assessment of size reduction as a preliminary step in the production of ethanol from lignocellulosic wastes. *Biol. Wastes*, *30*, 153–157.

Chang, V., Burr, B. & Holtzapple, M. (1997). Lime pretreatment of switchgrass. *Appl. Biochem. Biotechnol.*, 63–65(1), 3–19

Chang, V. S. & Holtzapple, M. T. (2000). Fundamental factors affecting biomass enzymatic reactivity. *Appl. Biochem. Biotechnol.*, 84-86, 5–37.

Chang, V. S., Nagwani, M. & Holtzapple, M. T. (1998). Lime pretreatment of crop residues bagasse and wheat straw. Appl. *Biochem. Biotechnol.*, *74*, 135–159.

Chang, V. S., Kaar, W. E., Burr, B. & Holtzapple, M. T. (2001a). Simultaneous saccharification and fermentation of lime-treated biomass. *Biotechnol. Lett.*, *23* (16), 1327–1333.

Chang, V., Nagwani, M., Kim, C. & Holtzapple, M. (2001b). Oxidative lime pretreatment of high-lignin biomass. *Appl. Biochem. Biotechnol.*, 94, 1–28.

Choi, C. H. & Oh, K. K. (2012). Application of a continuous twin screw-driven process for dilute acid pretreatment of rape straw. *Bioresour. Technol.*, *110*, 349–354.

Chundawat, S. P., Vismeh, R., Sharma, L. N., Humpula, J. F., da Costa Sousa, L., Chambliss, C. K., Jones, A. D., Balan, V. & Dale, B. E. (2010). Multifaceted characterization of cell wall decomposition products formed during ammonia fiber expansion (AFEX) and dilute acid based pretreatments. *Bioresour. Technol.*, *101*(21), 8429–8438.

Chundawat, S. P. S., Bals, B., Campbell, T., Sousa, L., Gao, D., Jin, M., Eranki, P., Garlock, R., Teymouri, F., Balan, V. & Dale, B. E. (2013). Primer on ammonia fiber expansion pretreatment. In: Wyman CE (ed) *Aqueous pretreatment of plant biomass for biological and chemical conversion to fuels and chemicals*. Wiley, New York, 169–200.

Ciesielski, P. N,. Wang, W., Chen, X., Vinzant, T. B., Tucker, M. P., Decker, S. R., Himmel, M. E., Johnson, D. K. & Donohoe, B. S. (2014). Effect of mechanical disruption on the effectiveness of three reactors used for dilute acid pretreatment of corn stover part 2: morphological and structural substrate analysis. *Biotechnol. Biofuels*, *7*(1), 47.

Couto, S. R. & Toca-Herrera, J. L. (2007). Laccase production at reactor scale by filamentous fungi. *Biotechnol. Adv.*, *25*, 558–569 .

Delgenés, J. P., Moletta, R. & Navarro, J. M. (1990). Acid-hydrolysis of wheat straw and process considerations for ethanol fermentation by Pichia Stipitis Y7124. *Process Biochemistry*, *25*, 132–135.

Delgenés, J. P., Penaud, V. & Moletta, R. (2002). Pretreatments for the enhancement of anaerobic digestion of solid wastes. In: *Biomethanization of the Organic Fraction of Municipal Solid Wastes*. IWA Publishing, 201–228 (Chap. 8).

Duque, A., Manzanares, P., Ballesteros, I., Negro, M. J., Oliva, J. M., Saez, F. & Ballesteros, M. (2014). Study of process configuration and catalyst concentration in integrated alkaline extrusion of barley straw for bioethanol production. *Fuel, 134*, 448–454.

Dwivedi, P., Vivekanand, V., Pareek, N., Sharma, A. & Singh, R. P. (2011). Co-cultivation of mutant Penicillium oxalicum SAUE-3.510 and Pleurotus ostreatus for simultaneous biosynthesis of xylanase and laccase under solid-state fermentation. *New Biotechnol., 28*(6), 616–626

Elander, R. T. (2013). Experimental pretreatment systems from laboratory to pilot scale. In: Wyman CE (ed) *Aqueous pretreatment of plant biomass for biological and chemical conversion to fuels and chemicals*. Wiley, UK, 417–450.

Gandi, J., Holtzapple, M. T., Ferrer, A., Byers, F. M., Turner, N. D., Nagwani, M. & Chang, S. (1997). Lime treatment of agricultural residues to improve rumen digestibility. *Anim. Feed Sci. Technol., 68*, 195–211.

Galbe, M. & Zacchi, G. (2007). Pretreatment of lignocellulosic materials for efficient bioethanol production. *Biofuels, 108*, 41–65.

Galhaup, C., Wagner, H., Hinterstoisser, B. & Haltrich, D. (2002). Increased production of laccase by the wood-degrading basidiomycete Trametes pubescens. *Enzyme Microb. Technol., 30*, 529–536.

Gao, M., Xu, F., Li, S., Ji, X., Chen, S. & Zhang, D. (2010). Effect of $SC-CO_2$ pretreatment in increasing rice straw biomass conversion. *Biosystems Engineering, 106*(4), 470–475.

García-Cubero, M. T., González-Benito, G., Indacoechea, I., Coca, M. & Bolado, S. (2009). Effect of ozonolysis pretreatment on enzymatic digest- ibility of wheat and rye straw. *Bioresour. Technol., 100*(4), 1608–1613

Georgieva, T. I., Hou, X., Hilstrøm, T. & Ahring, B. K. (2008). Enzymatic hydrolysis and ethanol fermentation of high dry matter wet- exploded wheat straw at low enzyme loading. *Appl. Biochem. Biotechnol., 148*(1–3), 35–44.

Gould, J. M. (1984). Alkaline peroxide delignification of agricultural residues to enhance enzymatic saccharification. *Biotechnol. Bioeng., 26*, 46–52.

Gregg, D. & Saddler, J. N. (1996). A techno-economic assessment of the pretreatment and fractionation steps of a biomass-to-ethanol process. *Appl. Biochem. Biotechnol.*, 711–727.

Grohmann, K., Torget, R. & Himmel, M. (1985). Optimization of dilute acid pretreatment of biomass. *Biotechnol. Bioeng. Symp., 15*, 59–80.

Hao, J. I., Tian, X., Song, F., He, X. B., Zhang, Z. H. & Zhang, P. (2006). Involvement of lignocellulolytic enzymes in the decomposition of leaf litter in a subtropical forest. *J. Eukaryot Microbiol., 53*, 193–198.

Haghighi Mood, S., Hossein Golfeshan, A., Tabatabaei, M., Salehi Jouzani, G., Najafi, G. H., Gholami, M. & Ardjmand, M. (2013). Lignocellulosic biomass to bioethanol, a comprehensive review with a focus on pretreatment. *Renewable and Sustainable Energy Reviews, 27*, 77–93.

Hatakka, A. (2001). Biodegradation of lignin. In: Hofrichter, M. Steinbuchel, A. (eds) Biopolymer. Biology, chemistry, biotechnology, application, vol 1 lignin, humic substance and coal. Wiley, New York.

He, Y., Pang, Y., Liu, Y., Li, X. & Wang, K. (2008). Physicochemical Characterization of Rice Straw Pretreated with Sodium Hydroxide in the Solid State for Enhancing Biogas Production. *Energ. Fuel.*, *22*, 2775–2781.

Hendriks, A. T. & Zeeman, G. (2009). Pretreatments to enhance the digestibility of lignocellulosic biomass. *Bioresour. Technol.*, *100*, 10–18.

Heinzkill, M., Bech, L., Halkier, T., Schneider, P. & Anke, T. (1998). Characterization of laccases and peroxidases from wood-rotting fungi (Family *Coprinaceae*). *Appl. Environ. Microbiol.*, *64*, 1601–1606.

Holtzapple, M. T., Jun, J. H., Ashok, G., Patibandla, S. L. & Dale, B. E. (1991). The ammonia freeze explosion (AFEX) process: a practical lignocellulose pretreatment. *Appl. Biochem. Biotechnol.*, (28/29), 59–74.

Hon, D. N. S. & Shiraishi, N. (2001). Wood and Cellulosic Chemistry, second ed. Dekker, New York.

Hu, Z. & Wen, Z. (2008). Enhancing enzymatic digestibility of switchgrass by microwave-assisted alkali pretreatment. *Biochemical Engineering Journal*, *38*, 369–378.

Hu, F., Jung, S. & Ragauskas, A. (2012). Pseudo-lignin formation and its impact on enzymatic hydrolysis. *Bioresour. Technol.*, *117*, 7–12.

Huijgen, W. J. J., Smit, A. T., Reith, J. H. & Uil, H. D. (2011). Catalytic organosolv fractionation of willow wood and wheat straw as pretreatment for enzymatic cellulose hydrolysis. *J. Chem. Technol. Biotechnol.*, *86*, 1428–1438.

Isroi, R. M., Syamsiah, S., Niklasson, C., Cahyanto, M. N., Ludquist, K. & Taherzadeh, M. J. (2011). Biological pretreatment of lignocelluloses with white-rot fungi and its applications: a review. *Bioresources*, *6*, 5224–5259.

Jacobsen, S.E . & Wyman, C. E. (2000). Cellulose and hemicellulose hydrolysis models for application to current and novel pretreatment processes. *Appl. Biochem. Biotechnol.*, 84–86, 81–96.

Jiang, W., Chang, S., Li, H., Oleskowicz-Popiel, P. & Xu, J. (2015). Liquid hot water pretreatment on different parts of cotton stalk to facilitate ethanol production. *Bioresour. Technol.*, *176*, 175–180.

Jouanin, L. & Lapierre, C. (eds), (2012). Lignins: biosynthesis, biodegradation and bioengineering. In: *Advances in botanical research*, vol 61. Academic Press, Elsevier, UK, Amsterdam.

Kaar, W. E. & Holtzapple, M. T. (2000). Using lime pretreatment to facilitate the enzymatic hydrolysis of corn stover. *Biomass Bioenerg.*, *18* (3), 189–199.

Kalyon, D. M. & Malik, M. (2007). An integrated approach for numerical analysis of coupled flow and heat transfer in co-rotating twin screw extruders. *Int. Polym. Process*, *22*, 293–302.

Kang, K. E., Jeong, G. T. & Park, D. H. (2012). Pretreatment of rapeseed straw by sodium hydroxide. *Bioprocess Bioprocess and Biosystems Engineering*, *35*(5), 705–713.

Karimi, K., Shafiei, M. & Kumar, R. (2013). Progress in physical and chemical pretreatment of lignocellulosic biomass. In: Gupta, V.K., Tuohy, M.G. (eds) *Biofuel technologies*. Springer, Berlin, 53–96.

Kim, K. H. & Hong, J. (2001). Supercritical CO_2 pretreatment of lignocellulose enhances enzymatic cellulose hydrolysis. *Bioresour. Technol.*, *77*, 139–144.

Kumakura, M., & Kaetsu, I. (1983). Effect of radiation pretreatment of bagasse on enzymatic and acid hydrolysis. *Biomass*, *3*, 199–208.

Kumakura, M. & Kaetsu, I. (1984). Pretreatment by radiation and acids of chaff and its effect on enzymatic hydrolysis of cellulose. *Agricultural Wastes*, *9*, 279–287.

Laser, M., Schulman, D., Allen, S. G., Lichwa, J., Antal, M. J. Jr. & Lynd, L. R. (2002). A comparison of liquid hot water and steam pretreatments of sugar cane bagasse for bioconversion to ethanol. *Bioresour. Technol.*, *81*(1), 33–44.

Li, Q., He, Y.-C., Xian, M., Jun, G., Xu, X., Yang, J.-M. & Li, L. Z. (2009). Improving enzymatic hydrolysis of wheat straw using ionic liquid 1-ethyl-3-methyl imidazolium diethyl phosphate pretreatment. *Bioresour. Technol. 100*, 3570–3575.

Linde, M., Jakobsson, E.-L., Galbe, M. & Zacchi, G. (2008). Steam pretreatment of dilute H_2SO_4-impregnated wheat straw and SSF with low yeast and enzyme loadings for bioethanol production. *Biomass Bioenerg.*, *32*, 326–332.

Lu, X. B., Xi, B., Zhang, Y. M. & Angelidaki, I. (2011). Microwave pretreatment of rape straw for bioethanol production: Focus on energy efficiency. *Bioresour. Technol.*, *102*, 7937–7940.

Luterbacher, J. S., Tester, J. W. & Walker, L. P. (2010). High-solids biphasic CO_2-H_2O pretreatment of lignocellulosic biomass. *Biotechnol. Bioeng.*, *107*(3), 451–460.

Lv, H., Ren, M., Zhang, M. & Chen, Y. (2013). Pretreatment of corn stover using supercritical CO_2 with water-ethanol as co-solvent. *Chin. J. Chem. Eng.*, *21*(5), 551—557.

Ma, H., Liu, W. W., Chen, X., Wu, Y. J. & Yu, Z. L. (2009). Enhanced enzymatic saccharification of rice straw by microwave pretreatment. *Bioresour. Technol.*, *100*, 1279–1284.

Ma, F., Yang, N., Xu, C., Yu, H., Wu, J. & Zhang, X. (2010). Combination of biological pretreatment with mild acid pretreatment for enzymatic hydrolysis and ethanol production from water hyacinth. *Bioresour. Technol.*, *101*, 9600–9604.

Martín, C., Marcet, M. & Thomsen, A. B. (2008). Comparison between wet oxidation and steam explosion as pretreatment methods for enzymatic hydrolysis of sugarcane bagasse. *BioResources*, *3*(3), 670–683.

Martin, E. M., Bunnell, K. A., Lau, C. S., Pelkki, M. H., Patterson, D. W., Clausen, E. C., Smith, J. A. & Carrier, D. J. (2011). Hot water and dilute acid pretreatment of high and low specific gravity Populus deltoides clones. *J. Ind. Microbiol. Biotechnol.*, *38*(2), 355–361.

McDermott, B., Chalmers, A. & Goodwin, J. (2001). Ultrasonication as a pre-treatment method for the enhancement of the psychrophilic anaerobic digestion of aquaculture effluents. *Environ. Technol.*, *22*, 823–830.

McMillan, J. D. (1992). Process for pretreating lignocellulosic biomass: a review. National Renewable Energy Lab, Golden.

McMillan, J. D. (1994). Pretreatment of lignocellulosic biomass. In: Himmel, M.E., Baker, J.O., Overend, R.P. (Eds.), *Enzymatic Conversion of Biomass for Fuels Production*. American Chemical Society, Washington, DC, 292–324.

Mes-Hartree, M., Dale, B. E. & Craig, W. K. (1988). Comparison of steam and ammonia pretreatment for enzymatic hydrolysis of cellulose. *Appl. Biochem. Biotechnol.*, *29*, 462–468.

Millet, M. A., Baker, A. J. 7 Scatter, L. D. (1976). Physical and chemical pretreatment for enhancing cellulose saccharification. *Biotech. Bioeng. Symp.*, *6*, 125–153

Mosier, N., Wyman, C., Dale, B., Elander, R., Lee, Y. Y., Holtzapple, M. & Ladisch, M. (2005). Features of promising technologies for pretreatment of lignocellulosic biomass. *Bioresour. Technol.*, *96* (6), 673–686.

Mora-Pale, M., Meli, L., Doherty, T. V., Linhardt, R. J. & Dordick, J. S. (2011). Room temperature ionic liquids as emerging solvents for the pretreatment of lignocellulosic biomass. *Biotechnol. Bioeng.*, *108*, 1229–1245.

Mtui, G. & Nakamura, Y. (2010). Lignocellulosic enzymes from Flavodon flavus, a fungus isolated from western Indian Ocean off the Coast of Dar es Salaam, Tanzania. *Afr. J. Biotechnol.*, *7*, 3066–3072.

Mtui, G. Y. S. (2010). Recent advances in pretreatment of lignocellulosic wastes and production of value added products. *Afr. J. Biotechnol. 8*, 1398–1415.

Narayanaswamy, N., Dheeran, P., Verma, S. & Kumar, S. (2013). Biological Pretreatment of Lignocellulosic Biomass for Enzymatic Saccharification. In: Z. Fang (ed.), *Pretreatment Techniques for Biofuels and Biorefineries*, Green Energy and Technology, DOI 10.1007/978-3-642-32735-3_1, Springer-Verlag, Berlin Heidelberg, 3-34. (Chap.1)

Ninomiya, K., Ohta, A., Omote, S., Ogino, C., Takahashi, K. & Shimizu, N. (2013). Combined use of completely bio-derived cholinium ionic liquids and ultrasound irradiation for the pretreatment of lignocellulosic material to enhance enzymatic saccharification. *Chem. Eng. J.*, 215–216, 811–818.

Oliveira, F. M. V., Pinheiro, I. O., Souto-Maior, A. M., Martin, C., Gonçalves, A. R. & Rocha, G. J. M. (2013). Industrial-scale steam explosion pretreatment of sugarcane straw for enzymatic hydrolysis of cellulose for production of second generation ethanol and value-added products. *Bioresour. Technol*, *130*, 168–173.

Olivier-Bourbigou, H., Magna, L. & Morvan, D. (2010). Ionic liquids and catalysis: recent progress from knowledge to applications, *Appl. Catal.* A, *373*, 1–56.

Ooshima, H., Aso, K. & Harano, Y. (1984). Microwave treatment of cellulosic materials for their enzymatic hydrolysis. *Biotechnol. Lett*, *6*, 289–294.

Overend, R. P. & Chornet, E. (1987). Fractionation of lignocellulosis by steam–aqueous pretreatments. *Philos. Trans. R. Soc. London*, A *321*, 523–536.

Palmowski, L. & Muller, J. (1999). Influence of the size reduction of organic waste on their anaerobic digestion. In: *II International Symposium on Anaerobic Digestion of Solid Waste*. Barcelona 15–17 June, 137–144.

Palmqvist, E. & Hahn-Hägerdal, B. (2000a). Fermentation of lignocellulosic hydrolysates. I: inhibition and detoxification. *Bioresour. Technol.*, *74*(1), 17–24.

Palmqvist, E. & Hahn-Hägerdal, B. (2000b). Fermentation of lignocellulosic hydrolysates. II: inhibitors and mechanisms of inhibition. *Bioresour. Technol.*, *74*(1), 25–33.

Park, J. Y., Shiroma, R., Al-Haq, M. I., Zhang, Y., Ike, M., Arai-Sanoh, Y., Ida, A., Kondo, M. & Tokuyasu, K. (2010). A novel lime pretreatment for subsequent bioethanol production from rice straw-Calcium capturing by carbonation (CaCCO) process. *Bioresour. Technol.*, *101*(17), 6805–6811

Peng, H., Luo, H., Jin, S., Li, H. & Xu, J. (2014). Improved bioethanol production from corn stover by alkali pretreatment with a novel pilot-scale continuous microwave irradiation reactor. *Biotechnol. Bioproc. E.*, *19*, 493-502.

Qiang, Z. & Thomsen, A. B. (2012). Effect of different wet oxidation pretreatment conditions on ethanol fermentation from corn stover. In: *Information Technology and Agricultral Engineering*. Springer-Verlag, Berlin Heidelberg, AISC, *134*, 953–958.

Rajendran, K. & Taherzaden, M. J. (2014). Pretreatment of Lignocellulosic Materials. In: *Bioprocessing of Renewable Resources to Commodity Bioproducts, First Edition*. John Wiley & Sons, Inc. Published, 43-75 (Chap. 3).

Reshamwala, S., Shawky, B. T. & Dale, B. E. (1995). Ethanol production from enzymatic hydrolysates of AFEX-treated coastal Bermuda grass and switchgrass. *Appl. Biochem. Biotechnol.*, 51/52, 43–55.

Sánchez, Ó. J. & Cardona, C. A. (2008). Trends in biotechnological production of fuel ethanol from different feedstocks. *Bioresour. Technol.*, *99*, 5270–5295.

Sánchez, C. (2009). Lignocellulosic residues: biodegradation and bioconversion by fungi. *Biotechnol. Adv.*, *27*, 185–194.

Sannigrahi, P., Hu, F., Pu, Y. & Ragauskas, A. (2012). A novel oxidative pre-treatment of Loblolly pine, Sweetgum, and Miscanthus by ozone. *J. Wood Chem. Technol.*, *32*(4), 361–375.

Schmidt, A. S., Mallon, S., Thomsen, A. B., Hvilsted, S., Lawther, J. M. (2002). Comparison of the chemical properties of wheat straw and beech fibers following alkaline wet oxidation and laccase treatments. *J. Wood Chem. Technol.*, *22*(1), 39–53.

Senturk-Ozer, S., Gevgilili, H. & Kalyon, D. M. (2011). Biomass pretreatment strategies via control of rheological behavior of biomass suspensions and reactive twin screw extrusion processing. *Bioresour. Technol.*, *102*(19), 9068–9075

Shafiei, M., Kumar, R. & Karimi K.(ed) (2015). Pretreatment of Lignocellulosic Biomass. In: *Lignocellulose-Based Bioproducts, Biofuel and Biorefinery Technologies* 1, DOI 10.1007/978-3-319-14033-9_3. Springer International Publishing, Switzerland, 85-154 (Chap. 3).

Sierra, R., Garcia, L. A. & Holtzapple, M. T. (2010). Selectivity and delignification kinetics for oxidative and nonoxidative lime pretreatment of poplar wood, part III: long-term. *Biotechnol. Prog.*, *26*(6), 1685–1694.

Sorensen, A., Teller, P. J., Hilstrøm, T. & Ahring, B. K. (2008). Hydrolysis of Miscanthus for bioethanol production using dilute acid presoaking combined with wet explosion pre-treatment and enzymatic treatment. *Bioresour. Technol.*, *99*(14), 6602–6607.

Sun, Y. & Cheng, J. (2002). Hydrolysis of lignocellulosic materials for ethanol production: a review. *Bioresour. Technol.*, *83*, 1–11.

Sun, S-L., Wen, J-L., Ma, M-G., Sun, R-C. 7 Jones, G. L. (2014). Structural features and antioxidant activities of degraded lignins from steam exploded bamboo stem. *Ind. Crop Prod.*, *56*, 128–136.

Sun, Z. Y., Tang, Y. Q., Iwanaga, T., Sho, T. & Kida, K. (2011). Production of fuel ethanol from bamboo by concentrated sulfuric acid hydrolysis followed by continuous ethanol fermentation. *Bioresour. Technol.*, *102*(23), 10929–10935.

Swatloski, R. P., Spear, S. K., Holbrey, J. D. & Rogers, R. D. (2002). Dissolution of cellose with ionic liquids. *Journal of the American Chemical Society*, *124*(18), 4974–4975.

Taherzadeh, M. J. & Karimi, K. (2007). Acid-based hydrolysis processes for ethanol from lignocellulosic materials: a review. *Bioresour. 2*, 472–499.

Taherzadeh, M. J. & Karimi, K. (2008). Pretreatment of lignocellulosic wastes to improve ethanol and biogas production: a review. *Int. J. Mol. Sci., 9*(9), 1621–1651.

Talebnia, F., Karakashev, D. & Angelidaki, I. (2010). Production of bioethanol from wheat straw: An overview on pretreatment, hydrolysis and fermentation. *Bioresour. Technol., 101*(13), 4744-4753.

Tan, H. T. & Lee, K. T. (2012). Understanding the impact of ionic liquid pretreatment on biomass and enzymatic hydrolysis. *Chem. Eng. J., 183* 448–458.

Tengborg, C., Stenberg, K., Galbe, M., Zacchi, G., Larsson, S., Palmqvist, E. & Hahn-Hägerdal, B. (1998). Comparison of SO_2 and H_2SO_4 impregnation of softwood prior to steam pretreatment on ethanol production. *Appl. Biochem. Biotechnol.*, 3–15.

Teymouri, F., Laureano-Perez, L., Alizadeh, H. & Dale, B. E. (2004). Ammonia fiber explosion treatment of corn stover. *Appl. Biochem. Biotechnol., 113*, 951–963.

Tian, X., Fang, Z. & Guo, F. (2012). Impact and prospective of fungal pre-treatment of lignocellulosic biomass for enzymatic hydrolysis. *Biofuels Bioprod. Biorefin.* 6, 335–350.

Tian, X., Fang, Z. (ed.), Xu, C. (2013). Status and perspective of organic solvent based pretreatment of lignocellulosic biomass for enzymatic saccharification. In: Fang, Z. (ed.), *Pretreatment Techniques for Biofuels and Biorefineries*, Green Energy and Technology, DOI 10.1007/978-3-642-32735-3_14, Springer-Verlag, Berlin Heidelberg, 309-336. (Chap. 14)

Tuor, U., Winterhalter, K. & Fiechter, A. (1995). Enzymes of white-rot fungi involved in lignin degradation and ecological determinants for wood decay. *J. Biotechnol., 41*, 1–17.

Vandenbossche, V., Brault, J., Vilarem, G., Hernández-Meléndez, O., Vivaldo-Lima, E., Hernández-Luna, M., Barzana, E., Duque, A., Manzanares, P., Ballesteros, M., Mata, J., Castellón, E. & Rigal, L. (2014). A new lignocellulosic biomass deconstruction process combining thermo-mechano chemical action and bio-catalytic enzymatic hydrolysis in a twin-screw extruder. *Ind. Crop Prod., 55*, 258–266.

Vidal, P. F. & Molinier, J. (1988). Ozonolysis of lignin–Improvement of in vitro digestibility of poplar sawdust. *Biomass, 16*(1), 1–17.

Walpot, J. (1986). Enzymatic hydrolysis of waste paper. *Conserv. Recycl., 9*, 127–136.

Wang, W., Chen, X., Donohoe, B. S., Ciesielski, P. N., Katahira, R., Kuhn, E. M., Kafle, K., Lee, C. M., Park, S., Kim, S. H., Tucker, M. P., Himmel, M. E. & Johnson, D. K. (2014). Effect of mechanical disruption on the effectiveness of three reactors used for dilute acid pretreatment of corn stover part 1: chemical and physical substrate analysis. *Biotechnol Biofuels, 7*, 57–69.

Wanderley, M. C. D. A., Martín, C. & Rocha, G. J. D. M. & Gouveia, E. R. (2013). Increase in ethanol production from sugarcane bagasse based on combined pretreatments and fed-batch enzymatic hydrolysis. *Bioresour. Technol., 128*(0), 448–453.

Wong, D. W. S. (2009). Structure and action mechanism of ligninolytic enzymes. *Appl. Biochem. Biotechnol., 157*, 174–209

Woiciechowski, A. L., de Souza Vandenberghe, L. P., Karp, S. G., Letti, L. A. J., de Carvalho, J. C., Medeiros A. B. P., Spier, M. R., Faraco, V., Soccol, V. T. & Soccol, C. R. (2013). The Pretreatment Step in Lignocellulosic Biomass Conversion: Current Systems and New Biological Systems. In: Faraco, V. (ed.), *Lignocellulose Conversion*,

DOI: 10.1007/978-3-642-37861-4_3, Springer-Verlag, Berlin Heidelberg, 39-64. (Chap. 3)

Wyman, C. E., Dale, B. E., Elander, R. T., Holtzapple, M., Ladisch, M. R. & Lee, Y. Y. (2005). Coordinated development of leading biomass pretreatment technologies. *Bioresour. Technol.*, *96*(18),1959–1966

Xiang, Q., Kim, J. S. & Lee, Y. Y. (2003). A comprehensive kinetic model for dilute-acid hydrolysis of cellulose. *Appl. Biochem. Biotechnol.*, 105–108, 337–352

Xiong, J., Ye, J., Liang, W. Z. & Fan, P. M. (2000). Influence of microwave on the ultrastructure of cellulose. *Journal of South China University Technology*, *28*, 84–89.

Xu, C., Ma, F., Zhang, X. 7 Chen, S. (2010). Biological pretreatment of corn stover by *Irpexlacteus* for enzymatic hydrolysis. *J. Agric. Food Chem.*, *58*, 10893–10898.

Xu, J. & Cheng, J. J. (2011). Pretreatment of switchgrass for sugar produc- tion with the combination of sodium hydroxide and lime. *Bioresour. Technol.*, *102*(4), 3861–3868.

Yang, B. & Wyman, C. E. (2004). Effect of xylan and lignin removal by batch and flow through pretreatment on the enzymatic digestibility of Corn Stover Cellulose. *Biotechnol. Bioeng.*, *86* (1), 88–95.

Yu, H., Guo, G., Zhang, X., Yan, K. & Xu, C. (2009). The effect of biological pretreatment with the selective white-rot fungus *Echinodontium taxodii* on enzymatic hydrolysis of softwoods and hardwoods. *Bioresour. Technol.*, *100*, 5170–5175.

Zhao, X., Cheng, K. & Liu, D. (2009). Organosolv pretreatment of ligno- cellulosic biomass for enzymatic hydrolysis. *Applied Microbiology and Biotechnology*, *82*(5), 815–827

Zhao, X., Zhang, L. & Liu, D. (2012). Biomass recalcitrance. Part II: fundamentals of different pre-treatments to increase the enzymatic digestibility of lignocellulose. *Biofuels, Bioprod. Biorefin.*, *6*, 561–579.

Zhang, S., Keshwani, D. R., Xu, Y. & Hanna, M. A. (2012). Alkali combined extrusion pretreatment of corn stover to enhance enzyme saccharification. *Ind. Crop Prod.*, *37*(1), 352–357.

Zheng, M. X., Li, X. J., Li, L. Q., Yang, X. J. & He, Y. F. (2009). Enhancing anaerobic biogasification of corn stover through wet state NaOH pretreatment. *Bioresour. Technol.*, *100*, 5140–5145.

Zheng, Y., Lin, H. M., Wen, J., Cao, N., Yu, X. & Tsao, G. T. (1995). Supercritical carbon dioxide explosion as a pretreatment for cellulose hydrolysis. *Biotechnol. Lett.*, *17*(8), 845–850.

Zheng, Y., Lin, H. M. & Tsao, G. T. (1998). Pretreatment for cellulose hydrolysis by carbon dioxide explosion. *Biotechnol. Prog.*, *14*, 890–896.

Zheng, Y., Zhao, J., Xu, F. & Li, Y. (2014). Pretreatment of lignocellulosic biomass for enhanced biogas production. *Prog. Energy Combust Sci.*, *42*, 35–53.

Zhu, S., Wu, Y., Chen, Q., Yu, Z., Wang, C., Jin, S., Ding, Y. & Wu, G. (2006). Dissolution of cellulose with ionic liquids and its application: A mini-review. *Green Chem.*, *8*(4), 325–327.

Zhu, J. Y., & Pan, X. J. (2010). Woody biomass pretreatment for cellulosic ethanol production: technology and energy consumption evaluation. *Bioresour. Technol.*, *101*(13), 4992–5002.

In: Gas Biofuels from Waste Biomass
Editor: Zhidan Liu

ISBN: 978-1-63483-192-5
© 2015 Nova Science Publishers, Inc.

Chapter 4

BIOGAS PRODUCTION FROM HIGH-SOLID ORGANIC WASTES AND ENERGY CROPS: DRY ANAEROBIC DIGESTION SYSTEM

*Valentine Nkongndem Nkemka and Xiying Hao**

Agriculture and Agri-Food Canada,
Lethbridge Research Centre, Alberta, Canada

ABSTRACT

Dry anaerobic digestion of solid waste has the advantage of small reactor design and simple operational prcoesses, which generally requires less labour and maintainance cost. This chapter reviews current dry anaerobic systems that are employed for the anaerobic digestion of high-solid organic wastes and energy crops. These systems include one- and two-stage systems, batch, sequential batch and continous processes. Pilot-scale demonstration of these systems are needed for continual optimization and its increase adpotion as a full-scale process. Enactment of legislation for the limit of landfilling or solid organic waste worldwide and government subcidies shall also act as a catalyst for the deployment of this technology.

Keywords: dry anaerobic digestion, leach bed reactor, organic loading rate, solid waste, UASB reactor

1. DRY ANAEROBIC DIGESTION CONCEPT

The anaerobic digestion of organic matter with a total solid (TS) content of > 15% is categorized under dry anaerobic digestion, while in a wet process the TS content ranges from 2-15% [1]. Dry anaerobic digestion has been used mainly in the digestion of municipal solid waste (MSW), energy crops, agricultural residues and manure that have high dry matter content (Table 1). It can be energetically and in some instances economically wasteful to

* Corresponding author: 1 403 317-2279; Fax: 1 403 317-3156; email: xiying.hao@agr.gc.ca.

digest these feed stocks in a wet process with a water content of 85-98%. Alternatively, dry digestion of these feedstocks can be more applicable and can lead to the design of small reactor volumes and reduce energy demand for heating, mixing and pumping, avoid foaming, and also reduces the handling cost during storage and spreading of the digestate as an amendment for improving soil quality [2]. In addition, little pretreatment by sorting is required [3]. However, clogging, channeling, uneven heating and distribution of microbial population over space and time, floating of feedstock in the leach bed are some of the disadvantages of dry anaerobic digestion over the wet process [2]. The current market share of dry anaerobic digestion, however, stands at 10% in Europe [4] and biogas production can become more competitive with the enactment of legislations that impose limit or fee for landfilling of organic wastes and the availability of government subsidies [5].

2. DRY DIGESTION REACTOR CONFIGURATIONS

Different configurations for dry anaerobic digestion exist and include one-stage versus two-stage processes, batch versus continuous processes or combination thereof.

2.1. One- or Two-Stage Anaerobic Digestions

This section focuses on dry anaerobic digestion configuration and has not address the conventional wet one- or two-stage digesters. In a one-stage anaerobic digestion, all four metabolic processes of hydrolysis, acidogenesis, acetogenesis and methanogenesis occur in a single reactor. While in a two-stage process, hydrolysis and acidogenesis occur primarily in the first reactor and acetogenesis and methanogenesis occur in the second reactor. Separation of these metabolic steps into two or more reactors is advantageous as optimum growth conditions for the hydrolyzing microbes differ significantly from the methanogens [13]. The hydrolyzing microbes have a higher growth rate and an optimum pH of 5.5 to 6.5, while the methanogens are slow growing and have an optimum pH of 6.8 to 7.6 [14]. The extent of separation of hydrolysis and methanogenesis in a two-stage anaerobic digestion process depends on the substrate and operation of the process [15]. In the digestion of easy hydrolysable organic matter, hydrolysis and methanogenesis can be easily separate with a high acidification of the first stage reactor and conversions of the soluble organics in the second stage methane reactor [15, 16]. However, hydrolysis and methane production can be predominant in the first reactor in the digestion of slowly degrading substrates or their combinations with a high buffer capacity [15].

2.2. Batch or Continuous Anaerobic Digestions

In batch anaerobic processes, the substrate is loaded in the beginning and allowed to digest for a long period of time until the methane production reaches a minimum. In continuous anaerobic processes, the substrate is continuously/semi continuously fed over time in order to maintain a constant high biogas production.

Leach Bed and Sequential Leach Bed Processes

Figure 1 shows the schematic leach bed and sequential leach bed processes. In a batch anaerobic process, the feedstock is loaded at the beginning and only removed at the end of the digestion when methane production reaches a preset minimum say after 60-90 days. The feeding strategy of the batch leach bed process makes it easy and less labor intensive to operate. This system also has a low investment and maintenance cost and is attractive as a farm-scale anaerobic digester and for developing countries.

At the beginning of the process, a fraction of the digested material is usually left over to act as inoculum source for the next batch, which speeds up the onset of methane production and reduces the acidification of the leach bed. However, blending digested material as inoculation with fresh material reduces the volumetric loading rate of the reactor. During the process, liquid (leachate) is recirculated and sprayed over the leach bed to improve contact between the bacterial biomass and the feedstock, and also to wash hydrolyzed soluble organics. The required mesophilic or thermophilic temperatures are obtained by heating the leachate collection reservoir, leachate pipeline or by heating the leach bed floor and wall. Batch anaerobic digestion is characterized by a period of low production (lag phase) followed by a period of high production (exponential phase) and then low production (stationary and decline phases). Hence, methane production in batch systems is not constant. This reactor configuration is suited for feedstocks with a high buffer capacity such as manure or when co-digesting feedstock. A methane yield of 0.19 L/g VS after 77 days of digestion of solid cow manure in a 1 L leach bed reactor has been previously reported [15] (Table 2). Gas-tight containers or garage leach bed or percolation reactors for the digestion of energy crops, MSW and agricultural residues are often used for full-scale operations such as those marketed by Bekon, Germany [1, 17]. The operation of several leach bed reactors with each starting at different times can lead to constant and high biogas production similar to continuous processes. Recirculation of the leachate of the old leach bed over the freshly started process is performed in order to accelerate the start-up phase, reduce handling and also to limit the recycle of old digested material which can increase the volumetric loading rate when compared to a leach bed digester [3]. An inoculum to substrate ratio of 20-80% w/w has been recommended for a quick start-up of the acidogenic phase in a leach bed reactor [18]. The effect of leachate exchange rate in the anaerobic digestion of food waste in a sequential leach bed process has been studied [19]. Their results showed that a higher leachate exchange rate between the start-up and mature reactor resulted in faster start-up time but there was no significant difference in the final methane yield. In addition, a pH of 6.5 and methane production rate > 0.5 L kg^{-1} VS day^{-1} were used as indicators to stop leachate exchange between start-up and mature reactors [19]. The Bekon full-scale leach bed process is made up of gas-tight rectangular concrete garage type building and also has at least three loading bays [1]. The required mesophilic temperature is maintained by floor and wall heating (www.bekon.eu). About 50% of digested material is mixed with fresh feed for the start-up of a new batch. The loading of the fermentation bays is performed in sequence to achieve a constant overall gas production. The leachate from the biomass drains into a collection tank from where it is sprinkled uniformly over the leach bed. A key safety feature of this system is the ventilation pump which purges residual methane from the container before opening/refilling the digester at the end of a batch and this also reduces methane emission of the digester. This safety feature is also used to create anaerobic atmosphere to accelerate the fermentation process. Low capital, maintenance, operation and labor costs are the advantages of this process [1].

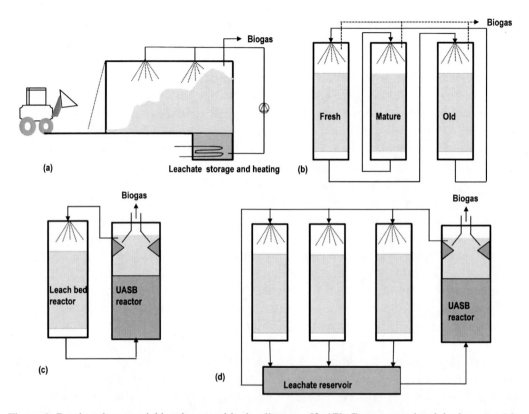

Figure 1. Batch and sequential batch anaerobic dry digesters [2, 17]. Garage-type leach bed reactor (a), sequential leach bed reactor (b), two-stage reactor (leach bed connected to a high rate methane reactor) (c), two-stage reactor (multiple leach bed connected to a high rate methane reactor).

Table 1. Characteristics of feedstocks used in dry anaerobic digestion

% of TS	MSW	Maize	Grass silage	Manure
TS (%)	21.1	18.0-52.9	25.5-30.7	1.6-32.2
VS (%)	17.4	17.4-50.1	24.1-28.4	1.2-29.5
C/N	24.6	24.2-52.1	26.0-26.7	18.9
Cellulose (% of TS)	-	19.3-37.3	34.3	3.8-42.9
Hemicellulose (% of TS)	-	25.3-38.0	29.6	1.0-22.2
Lignin (% of TS)	-	4.3-7.5	8.6	0.4-11.9
Crude protein (% of TS)	8.1	5.9-10.1	9.5-10.1	-
Crude fat (% of TS)	-	1.2-2.6	3.3	1.1-12.1
pH	3.9	-	4.3	7.3
References	(6)	(7)	(8, 9)	(10-12)

Two-Stage and Sequential Two-Stage Processes

Operation of a leach bed or several leach bed reactors in sequence can also be connected to a methane high-rate reactor. This two-stage coupled system is suitable for easily hydrolysable feedstock (energy crops and MSW), which can rapidly acidify the leach bed reactor, inhibiting methanogenesis. Coupling of a second methane reactor will rapidly convert the produced acidic leachate into CH_4 and CO_2 and thereby, improving the overall process stability. Examples of high-rate methane reactor commonly used are the upflow anaerobic sludge blanket (UASB) reactor or the methane filter or biofilm reactor (Figure 1). These reactors can retained a high biomass concentration in the dense granules or by microbial attachment on carriers thus, limiting the wash out of active bacterial and methanogenic biomass. Hence, high-rate methane reactor can sustain high organic loading rate (OLR) of 10 to 15 g COD/L.d [20-22]. Operation of several leach bed reactors can provide enough leachate to fully maximize the full organic loading capacity of a methane high-rate reactor. Biopercolate, WEHRLE Umwelt GmbH [3], is an example of a full-scale two-stage process that consists of a leach bed reactor connected to a methane filter. In this digester system, the leach bed reactor is equipped with a slow rotating horizontal mixer in the middle of the leach bed, which prevents clogging and channeling. Also, the leach bed reactor is slightly aerated at its base and after 2-3 days of hydrolysis, the produced leachate is circulated for methane production in the methane filter while the partially digested substrate is further decomposed in a tunnel composter. Another method for reducing clogging and channeling in the leach bed is by digestion of the feedstock together with a bulking agent such as wood chips. In a previous pilot-scale study of the anaerobic digestion of grass silage in 6 leach bed reactors (6× 17 L) connected to a UASB reactor (34.1 L), an increase in methane yield from 0.310 to 0.341L/g VS and a VS reduction of 70.5-75.5% was obtained when an extra pump was introduced to increase the recirculated volume of the leachate [23].

2.3. Continuous Dry Digestion Processes

In a continuous dry anaerobic digestion process, the digester is continuously or semi-continuously fed daily to maintain a constant biogas production. Feeding of feedstocks with a high dry matter content is usually performed with different pump types such as a flush system, piston and screw pumps [2]. During the process, the separated liquid effluent is used to dilute fresh feedstock while the solid could be further composted or used for combined heat and power production. Common continuous dry digesters used in anaerobic digestion include, vertical leach bed reactors, horizontal tubular plug flow reactors, inclined plug flow reactors and hybrid plug flow reactors (Figure 2).

Examples of commercialized full-scale dry digesters that have a large market share in Europe include Dranco, Volarga and Kompogas processes. The Dranco process is a plug flow vertical reactor with a conical base. The recycle of digested and fresh material is in a ratio of 6/1, which is mixed outside the reactor. The mix is then steamed to attain the required thermophilic temperature before feeding at the top of the reactor. The separated solid fraction of the digestate is further stabilized aerobically for use in agriculture. The application of high and sustainable organic loading rate of 15 g VS/L.d is an interesting feature of the Dranco process. The Volarga process also uses a vertical column plug flow reactor and operating at either mesophilic or thermophilic temperatures and at a retention time of 20 days. The reactor has a vertical wall in the middle and the substrate is fed through the bottom and moves around the wall and exits on the opposite side of the reactor (Figure 2d).

Compressed biogas is sparged at the base for mixing in the reactor thus, avoiding an inoculation loop as in the Dranco and Kompogas processes. New hydride designs such as HormozMehr® similar to the Volarga process are under pilot-scale evaluation [26]. The kompogas process uses a horizontal column unlike in the former processes. The substrate is feed at one end and slow moving impellers move the material gradually towards opposite end as they are digested. Solid-liquid separation of the effluent is performed to obtain a liquid that is recycled as inoculation for fresh material, while the solids are further stabilized by composting. The process is operated at a retention time of 20 to 25 days and under thermophilic temperature. A methane yield of 0.325 L/g VS was reported in the anaerobic co-digestion of cotton stalk and swine manure using HormozMehr® hybrid plug flow design [26]. Inclined plug flow continuous dry digesters are also being tested at pilot-scale. This system avoids the need for slow moving mixers and relies on movement due to gravity and during feeding of the reactor. A high methane yield of 0.327 L/g VS was obtained under thermophilic condition in the digestion of food waste with a C/N ratio of 27 and a retention time of 19 days.

It is difficult to make a fair comparison of the different dry digestion processes due to substrate variability, operational conditions (temperature, OLR and retention time) and variability in the testing methods and instruments. However, plug flow dry digesters can accommodate high OLRs when compared to wet processes (Table 2).

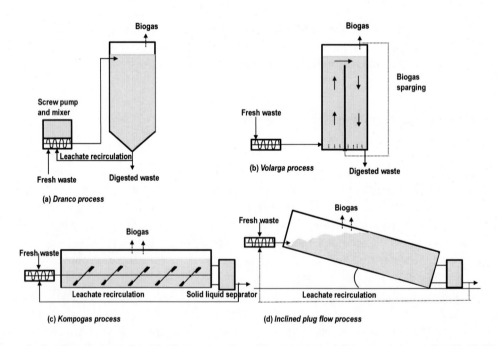

Figure 2. Continuous dry anaerobic digesters: a, b and d were adopted from previous studies [2, 3]. The inclined plug flow system was adopted according to earlier studies [24, 25].

3. FUTURE WORK AND RECOMMENDATIONS

Development of new dry anaerobic digestion systems should take into consideration cost, simplicity and maintenance of operation, safety, treatment of a wide variety of feedstocks and environmental friendly systems with reduced greenhouse gas emissions and leakage of nutrients. More research at pilot scale is needed to determine and optimize other operating parameters, which affect the dry digestion process. Finally, enactment of legislation of the ban or limit of land filling of solid organic wastes in countries worldwide and also the provision of incentives and subsidies will lead to the continual development of dry anaerobic digestion.

Table 2. Process performance of dry anaerobic digestion systems

Process type	Substrate	Reactor volume	Temperature (°C)	HRT (days)	OLR (g VS/L.d)	CH$_4$ yield (L/gVS)	Reference
One-stage CSTR	Energy crops	2.2 m^3	37	50	3.4	0.38	[27]
One-stage Batch leach bed	Cow manure	1 L	37	77	-	0.14	[15]
Sequential leach bed process	Food waste	12×120 L	37	73	-	0.23	[19]
Kompogas plug flow	Vegetable waste	15 m^3	55	21	-	0.46	[28]
Continuous one-stage tubular reactor	Food waste	60 L	Mesophilic	20-60	3-10	0.502	[6]
Hydride plug flow reactor	cotton stalk/pig manure	-	26-31	32	-	0.325	[29]
Sequential leach bed-UASB reactor	Grass silage	6×(17)-31.4 L		30-42	6.4[a]	0.31-0.34	[23]
Inclined plug flow reactor	Food waste	0.69 m^3	55	19	7-10	0.327	[24]
Batch leach bed- methane filter	MSW	10-2.8 m^3	37	67	1-3[a]	0.98	[30]
Continuous leach bed-methane filter	Maize silage	35 L-22 L	38	-	4.5	0.330	[31]

[a] CH4 yield in g COD/L.d
HRT: hydraulic retention time.

REFERENCES

[1] Li, Y; Park, SY; Zhu, J. Solid-state anaerobic digestion for methane production from organic waste. Renewable Sustainable *Energy Rev.*, 2011, 15(1), 821-6.

[2] Nizami, A-S; Murphy, JD. What type of digester configurations should be employed to produce biomethane from grass silage?. *Renewable Sustainable Energy Rev.*, 2010, 14, 1558-68.

[3] Rapport, J; Zhang, R; Jenkins, BM; Williams, RB. Current anaerobic digestion technologies used for treatment of municipal organic solid waste. *Califonia*, USA, 2008.

[4] Weiland, P. Biogas production: current state and perspectives. *Appl Microbiol Biotechnol.*, 2010, 85 849-60.

[5] Lantz, M; Svensson, M; Björnsson, L; Börjesson, P. The prospects for an expansion of biogas systems in Sweden—Incentives, barriers and potentials. *Energy Policy.*, 2007, 35(3), 1830-43.

[6] Cho, S-K; Im, W-T; Kim, D-H; Kim, M-H; Shin, H-S; Oh, S-E. Dry anaerobic digestion of food waste under mesophilic conditions: Performance and methanogenic community analysis. *Bioresour Technol.*, 2013, 131(0), 210-7.

[7] Amon, T; Amon, B; Kryvoruchko, V; Machmüller, A; Hopfner-Sixt, K; Bodiroza, V; et al. Methane production through anaerobic digestion of various energy crops grown in sustainable crop rotations. *Bioresour Technol.*, 2007, 98(17), 3204-12.

[8] Xie, S; Frost, JP; Lawlor, PG; Wu, G; Zhan, X. Effects of thermo-chemical pre-treatment of grass silage on methane production by anaerobic digestion. *Bioresour Technol.*, 2011, 102(19), 8748-55.

[9] Abdul-Sattar, N; D, MJ. Optimizing the Operation of a Two-Phase Anaerobic Digestion System Digesting Grass Silage. *Environ Sci Technol.*, 2011, 45(17), 7561-9.

[10] Triolo, JM; Sommer, SG; Møller, HB; Weisbjerg, MR; Jiang, XY. A new algorithm to characterize biodegradability of biomass during anaerobic digestion: Influence of lignin concentration on methane production potential. *Bioresour Technol.*, 2011, 102(20), 9395-402.

[11] Kalamaras, SD; Kotsopoulos, TA. Anaerobic co-digestion of cattle manure and alternative crops for the substitution of maize in South Europe. *Bioresour Technol.*, 2014, 172(0), 68-75.

[12] Hills, DJ. Effects of carbon: Nitrogen ratio on anaerobic digestion of dairy manure. *Agric Wastes.*, 1979, 1(4), 267-78.

[13] Ghosh, S; Henry, MP; Sajjad, A; Mensinger, MC; Arora, JL. Pilot-scale gasification of municipal solid wastes by high-rate and two-phase anaerobic digestion (TPAD). *Water Sci Technol.*, 2000, 41(3), 101-10.

[14] Jha, AK; Li, J; Nies, L; Zhang, L. Research advances in dry anaerobic digestion process of solid organic wastes. *Afr J Biotechnol.*, 2011, 10 (65), 14242-53.

[15] Nkemka, VN; Arenales-Rivera, J; Murto, M. Two-stage dry anaerobic digestion of beach cast seaweed and its codigestion with cow manure. *J Waste Manage.*, 2014, 2014, 9.

[16] Nkemka, VN; Murto, M. Two-stage anaerobic dry digestion of blue mussel and reed. *Renewable Energy.*, 2013, 50(0), 359-64.

[17] Nordberg, U; Nordberg, Å. Torrötning -kunskapssammanställning och bedömning av utvecklingsbehov. 2007.

[18] Xu, SY; Karthikeyan, OP; Selvam, A; Wong, JWC. Effect of inoculum to substrate ratio on the hydrolysis and acidification of food waste in leach bed reactor. *Bioresour Technol.*, 2012, 126(0), 425-30.

[19] Dearman, B; Bentham, RH. Anaerobic digestion of food waste: Comparing leachate exchange rates in sequential batch systems digesting food waste and biosolids. *Waste Manage.*, 2007, 27(12), 1792-9.

[20] Nkemka, VN; Murto, M. Biogas production from wheat straw in batch and UASB reactors: The roles of pretreatment and seaweed hydrolysate as a co-substrate. *Bioresour Technol.*, 2013, 128(0), 164-72.

[21] Lettinga, G; Hulshoff Pol, LW; Koster, IW; Wiegant, W; Zeeuw, WJd; Rinzema, A; et al. High rate anaerobic waste water treatment using the UASB-reactor under a wide range of temperature conditions. *Biotechnol Genet Engineering.*, 1984, 2, 253-84.

[22] Van der Merwe, M; Britz, TJ. Anaerobic digestion of baker's yeast factory effluent using an anaerobic filter and a hybrid digester. *Bioresour Technol.*, 1993, 43(2), 169-74.

[23] Nizami, A-S; Murphy, JD. Optimizing the operation of a two-phase anaerobic digestion system digesting grass silage. *Environ Sci Technol.* 2011, 45(17), 7561-9.

[24] Zeshan Karthikeyan, OP; Visvanathan, C. Effect of C/N ratio and ammonia-N accumulation in a pilot-scale thermophilic dry anaerobic digester. *Bioresour Technol.*, 2012, 113, 294-302.

[25] Karthikeyan, OP; Visvanathan, C. Bio-energy recovery from high-solid organic substrates by dry anaerobic bio-conversion processes: a review. *Rev Environ Sci Biotechnol.*, 2013, 12(3), 257-84.

[26] Adl, M; Sheng, K; Xia, Y; Gharibi, A; Chen, X. Examining a hybrid plug-flow pilot reactor for anaerobic digestion of farm-based biodegradable solids. *International Journal of Environmental Research.* 2011, 6(1), 335-44.

[27] Nges, IA; Björn, A; Björnsson, L. Stable operation during pilot-scale anaerobic digestion of nutrient-supplemented maize/sugar beet silage. *Bioresour Technol.*, 2012, 118(0), 445-54.

[28] Wellinger, A; Wyder, K; Metzler, A. Kompogas-a new system for the anaerobic treatment of source separated waste. Water science and technology: *a journal of the International Association on Water Pollution Research.* 1993, 27(2), 153-8.

[29] Adl, M; Sheng, K; Xia, Y; Gharibi, A; Chen, X. Examining a hybrid plug-flow pilot reactor for anaerobic digestion of farm-based biodegradable solids. *Int J Environ Res.*, 2011, 6(1), 335-44.

[30] Murto, M; Björnsson, L; Rosqvist, H; Bohn, I. Evaluating the biogas potential of the dry fraction from pretreatment of food waste from households. *Waste Manage.*, 2013, 33(5), 1282-9.

[31] Linke, B; Rodríguez-Abalde, Á; Jost, C; Krieg, A. Performance of a novel two-phase continuously fed leach bed reactor for demand-based biogas production from maize silage. *Bioresour Technol.*, 2015, 177(0), 34-40.

In: Gas Biofuels from Waste Biomass
Editor: Zhidan Liu

ISBN: 978-1-63483-192-5
© 2015 Nova Science Publishers, Inc.

Chapter 5

BIOGAS PRODUCTION FROM HIGH-SOLID ORGANIC BIOWASTES

Genevieve Wamaitha Kimari, Weizhong Jiang[*] and Kaiqiang Zhang

College of Water resource and Civil Engineering,
China Agricultural University, Beijing, China

ABSTRACT

This chapter covers the high solid biowaste feedstock resources and the forms in which they are available in the world. Wastes from agriculture, forest biomass, and municipalities and their specific sources have been highlighted. It also gives an overview of the major characteristics of particular substrates that influence the potential biomethane production of feedstock such as their physical characteristics.

Common bioreactor systems for anaerobic digestion have been discussed at length, indicating their working principles and recent advances in technology to make them more effective or suitable for specific feedstock. In this chapter, continuously stirred tank reactors (CSTR), anaerobic leach-bed reactors (ALBR), upflow anaerobic digestion of solid-state biowastes reactors (USR), anaerobic fluidized bed reactors (AFBR), batch stirring tank reactors (BSTR) and rotational drum fermentation systems (RDFS) have been discussed.

In order to realize the enhancement of biogas production, some technical methods have been used to accelerate hydrolysis and acidification processes. The last part of this section discusses process configuration and important parameters that should be controlled in anaerobic bioreactors. Special consideration should be given to temperature, pH, appropriate pretreatment of substrates and inhibition management, which employs some physical methods such as mineral clay particle adsorption and recirculation of leachate.

Keywords: high-solid biowaste, potential biomethane production, bioreactor systems, enhancement of biogas production, pretreatment, inhibition management

[*]Corresponding author: E.mail: jiangwz@cau.edu.cn.

1. BACKGROUND

Biowaste feedstocks can be broadly classified into dilute low solid wastes, intermediate solid wastes, and high solid wastes with < 1%, 5-10%, and > 10% solids concentration in the feedstock respectively (David P. Chynoweth, 2001). The focus of this chapter is on high solid concentration feedstock.

In recent years, there has been a lot of public outcry against use of coal generated energy propelled by the global focus on looming climate change, ever-increasing fuel prices, and the opportunity cost presented by the use of organic substrate for energy production.

Solid-state organic biowastes is an attractive resource for biogas production due to the fact that it possesses very high volumetric energy potential with respect to general wastewater and is also distributed widely in agricultural and industrial fields. Biogas production from high-solid organic biowastes is a promising energy source in the short- and long-term future. It is also the most important option in regard to technological, industrial, and environmental perspectives both for the present and future sustainable development. High solid anaerobic digestion is an advantageous management option since it involves smaller space requirements, a high volumetric organic loading rate, lower water and energy consumption, and easy handling of digested waste (E. Aymerich, 2013).

In practice, biogas production from high-solid organic biowastes has some sound benefits, which include: decreasing the wastes volume to be treated, lowering energy demands for maintaining mesophilic or thermophilic conditions in reactors, and working toward a zero-discharge of sludge and effluent.

2. RESOURCES, CHARACTERISTICS, AND POTENTIAL

Several ways of waste utilisation have been developed in recent years. Thus, waste, in particular organic waste, is no longer regarded as a material to eliminate but rather a source to explore. Effective waste management in the local, regional and global scale with overall aim of protecting human safety and sanitation standards as well as environmental protection requires reliable and complete statistical data on:

1. Waste sources
2. Amounts generated
3. Qualitative and quantitative structure, and properties
4. Environmental behavior and changes in waste streams as a function of time

Biowaste primarily constitutes:

- biowaste contents of carbohydrates, both starchy and lignocellulosic based which are potential fermentation substrates for bioenergy carriers, chemicals, and food/feed ingredient (e.g., amino acids),
- biowaste ash contents, which are potential plant nutrients as in fertilizer, and
- biowaste proteins, which are important ingredients in food and feed applications.

Anaerobic digestion (AD) is a complex bioconversion process that can produce abundant benefits for treating organic wastes, such as recovering energy in the form of biogas, producing organic fertilizer, and controlling greenhouse gas emission (Rapport J.L., 2011). The range of anaerobic digestion substrate is applicable to almost all organic material (Rilling, undated). The properties of these substrates greatly influence the potential for biogas formation, and the choice of suitable management strategy is dependent on the wastes properties such as structure and humidity (Rilling, undated). Different feedstocks require diverse anaerobic digestion technologies and some may even require pretreatment. The total solids (TS) concentration of the waste influences the pH, temperature and effectiveness of the microorganisms in the decomposition process of anaerobic digestion. TS and Volatile Solids (VS) are important parameters used to characterize feedstock's organic strength. VS especially indicates the fraction of the solid material that may be transformed into biogas (Wilkie, 2013). In regards to the cell structure presence of high amounts of lignin (in relation to cellulose and hemicelluloses) usually indicates low biodegradability of the feedstock and may cause need for pretreatment before anaerobic digestion (Richard, 1996). Bio-methane potential refers to an estimate of methane that could be produced from a particular feedstock by anaerobic digestion at optimum temperatures and conditions. This value may be used to compare how much available/potential methane yields a substrate has and the actual produced amounts which can then evaluate the performance or suitability of a feedstock in anaerobic digestion and also give insight on the applied anaerobic digestion system.

The following is an overview of the major biowastes that have been recognized in various researches as sources of biogas through anaerobic digestion.

For the purpose of this section the biowastes have been categorized into

1. Agricultural based wastes i.e., crop residue, food wastes, and animal manure
2. Forest wastes i.e., leaves, wood processing by products
3. Municipal Wastes i.e., organic fraction of MSW and Sewage sludge

2.1. Agricultural Wastes

Agriculture sector has two main sources of biowaste: wastes that result from livestock husbandry and those that arise from crop growing and utilization.

Livestock Waste

Animal waste includes voided waste from livestock and poultry, wastewater, feedlot runoff, silage juices, bedding, and feed.

Factors which contribute to the methane potential of manures are the animal species, breed, and growth stage of the animals, feed, amount, and type of bedding and also any degradation processes which may take place during storage. Different species have varying manure compositions; Cattle manure is largely composed of carbohydrates compared to pig slurry which is predominantly proteins; this leads to higher methane content in biogas produced from pig slurry (Dong Renjie, 2013). The total nitrogen contents of fresh goat manure (1.01%) and chicken manure (1.03%) are significantly higher than those of dairy manure (0.35%) and swine manure (0.24%) (Tong Zhang, 2013). Aspects such as the ammonia concentration in the livestock biowaste may also contribute to the biogas production potential.

Crop

When determining the suitability of using energy crops for biogas production a whole energy crop production life cycle assessment approach is required. Considerations such as seasonal vs. year round availability, labour and mechanization, land, soil and climate requirements, need for crop specific processing techniques, substrate "quality," among others should be considered (Dong Renjie, 2013, Thomas Amon, 2007). Crop yields are also varied depending on local environment and growth conditions. Ethical concerns about use of whole food crops as raw materials prompted research efforts toward inedible plants (and parts of crops). Aside from being an environmentally friendly process, agricultural residues help to avoid reliance on forest woody biomass and thus reduce deforestation. Unlike trees, crop residues are characterized by a short-harvest rotation that renders them more consistently available. Over the years with development of novel estimation technologies, the amount of available crop residues have varied but the overall consensus is that there is an abundant production majorly as a response to the increase in population and the intensive agriculture production technologies applied. According to FAOSTAT (FAO, 2014) the top five crops produced in the world in 2013 were wheat, rice, maize, sugarcane, and root crops with a total world production of 713.2, 745.7, 1016.7, 1877.1, and 835.9 million tons respectively. The average data of crop produced in a region can be used to estimate the potential of available crop residues that are applicable in anaerobic digestion.

Table 1. Physical Characteristics of Livestock Manure

Substrate	TS %	VS % (% of TS)
Cattle slurry	10	80
Pig slurry	6	80
Cattle Dung	25	80
Poultry manure	40	75
Horse manure w/o straw	28	75
Goat manure	33.65	82.21

Data compiled from reviews and research work by (Tong Zhang, 2013, Dong Renjie, 2013).

Table 2. Biomethane potential and methane yields of Livestock manure

Substrate	Biogas Methane yield (Nm^3/t FM)
Cattle slurry	14
Pig slurry	17
Cattle Dung	44
Poultry manure	90
Horse manure w/o straw	35

Data compiled from review and research work by (Dong Renjie, 2013)

Rice and Wheat

Residues from rice and wheat crops are straw, bran, and husk. Rice straw is the stems left over after the fruiting stalks are threshed and harvested. Bran refers to the outer layer and the germ of the rice or wheat grain. It is a by-product of whitening (removing all or part of the bran layer and germ) and\or polishing. Rice husk is the outermost layer of protection encasing a rice grain with a yellow colour (Bartali ElHoussine, 2014). Highest wheat production was in China India, and America each producing 12.7, 93.5, and 58 million tons while China India and Indonesia had the highest rice production in the world at 205, 159.2, and 71.3 million tons each(FAO, 2014).

Maize

These residues consist of the leaves and stalks of maize plants left in a field after harvest. Maize has a high energy yield per hectare, easy to digest in biogas plant, un-demanding, can be used as a whole plant silage or just the corn cob. Corn cob has higher energy density than whole plant silage.

Table 3. Physical Characteristics of selected Agricultural crops and agricultural processing wastes

Substrate	TS %	VS % (% of TS)
Maize silage	33	95
WCC silage	33	9
Green rye silage	25	90
Cereal grains	87	97
Grass silage	35	90
Sugar beet	23	90
Fodder beet	16	90
Sunflower silage	25	90
Sweet sorghum	22	91
Green rye	25	88
Wheat straw	81.08	90.29
Corn straw	81.74	91.04
Rice straw	77.93	94.23
Spent grains	23	75
Cereal vinasse	6	94
Potato vinasse	6	85
Fruit pomace	2.5	95
Rapeseed cake	92	87
Potato pulp	13	90
Potato juice	3.7	73
Pressed sugar beet pulp	24	95
Molasses	85	88
Apple pomace	35	88
Grape pomace	45	85
Sweet potato	35.5	97.6
Papaya	10.7	92.4
Straw	94.9	94.8

Data compiled from reviews and research work by (Tong Zhang, 2013, Dong Renjie, 2013, Xumeng Ge, 2014).

When using the entire maize plant, it is chopped, compacted, and sealed with an airtight plastic for about 12 weeks (Dong Renjie, 2013). Harvesting time can significantly affect the biogas yield of plants. According FAO the largest producers of maize in 2013 were America, China, and Brazil each producing 353.7, 217.8, and 80.5 million tons(FAO, 2014).

Table 4. Biomethane potential of selected Agricultural crops and agricultural processing wastes

Substrate	Biogas Potential (lg^{-1} VS)
Kelp (macrocystis)	0.39 - 0.41
Sorghum	0.26 - 0.39
Sargassum	0.26 - 0.38
Nappier grass	0.19 - 0.34
Poplar	0.23 - 0.32
Water Hyacinth	0.19 - 0.32
Sugarcane	0.23 - 0.30
Willow	0.13 - 0.30
Laminaria	0.26 - 0.28

Data compiled from reviews and research work by (David P. Chynoweth, 2001)

Table 5. Biogas methane yields of selected Agricultural crops and agricultural processing wastes

Substrate	Methane yield (Nm^3/t FM)
Maize silage	106
WCC silage	105
Green rye silage	79
Cereal grains	329
Grass silage	98
Sugar beet	72
Fodder beet	50
Sunflower silage	68
Sweet sorghum	58
Green rye	70
Spent grains	70
Cereal vinasse	22
Potato vinasse	18
Fruit pomace	9
Raw glycerol	147
Rapeseed cake	317
Potato pulp	47
Potato juice	30
Pressed sugar beet pulp	49
Molasses	229
Apple pomace	100
Grape pomace	176
Winter rye	0.36**

Substrate	Methane yield (Nm³/t FM)
Oilsedd rape	0.42**
Faba bean straw	0.441**
Maize (whole crop silage)	0.390**
Winter wheat straw	0.189**
Summer barley straw	0.189**
Sugar beet leaves	0.210**
Sunflower (whole crop silage)	0.300**
Maize early harvest	0.313 – 0.366**
Maize mid harvest	0.302 – 0.326**
Maize late harvest	0.268 – 0.287**

Data compiled from reviews and research work by (Dong Renjie, 2013, Alastair J. Ward, 2008)
** units in lg^{-1} VS.

Other crops

The leaves, stems, pseudo-stems, and discarded fruit of other crops such as bananas, sugarcane, root crops and tubers among others have been used for anaerobic digestion. Sugarcane production was recorded at 739.3, 341.2, and 126.1 million tons by Brazil, India, and China respectively, while China, Nigeria, and India were recorded o have the highest production of root crops with 174.5, 100, and 53.7 million tons respectively (FAO, 2014).

Table 6. Carbohydrate composition of various Agricultural crops and agricultural processing wastes

Feedstock	Carbohydrate composition (%dry weight)		
	Cellulose	Hemicellulose	Lignin
Barley hull	34	36	19
Barley straw	33.8 – 43	24 – 37.2	6.3 – 19.3
Corn cob	32.3 - 45.6	39.8	6.7 - 13.9
Corn stover	35.1 -39.5	20.7 -24.6	11.0 – 19.1
Cotton	85 -95	5 - 15	0
Cotton stalk	31	11	30
Coffee pulp	33.7 – 36.9	44.2 – 47.5	15.6 – 19.1
Rice straw	29.2 – 34.7	23 – 25.9	17 - 19
Rice husk	28.7 – 35.6	11.96 – 29.3	15.4 - 20
Wheat straw	35 - 39	22 – 30	12 - 16
Wheat bran	10.5 – 14.8	35.5 – 39 .2	8.3 – 12. 5
Sugarcane bagasse	25 - 45	28 - 32	15 - 25
Sugarcane tops	35	32	14
Winter rye	29 - 30	22 - 26	16 - 1
Oil seed rape	27.3	20.5	14.2
Oat straw	31 - 35	20 - 26	10 - 15
Nut shells	25 - 30	22 - 28	30 - 40
Sorghum straw	32 - 35	24 - 27	15 - 21
Tamarind kernel powder	10 -15	55 - 65	-
Water hyacinth	18.2 - 22.1	48.7 - 50.1	3.5 - 5.4
Agricultural residues	25 – 30	37 – 50	5 – 15
Barley hull	33.6	37.2	19.3
Barley straw	33.8	21.9	13.8

Table 6. (Continued)

Feedstock	Carbohydrate composition (%dry weight)		
	Cellulose	Hemicellulose	Lignin
Corn cobs	33.7	31.9	6.1
Corn stover	38.3	25.8	17.4
Cotton stalks	14.4	14.4	21.5
Wheat straw	30.2	22.3	17
Rice straw	31.1	22.3	13.3
Rye straw	30.9	21.5	25.3
Oat straw	39.4	27.1	17.5
Soya stalks	34.5	24.8	9.8
Sunflower stalks	42.1	29.7	13.4
Sugarcane bagasse	43.1	31.1	11.4
Sweet sorghum bagasse	27.3	14.5	14.3
Forage sorghum	35.6	20.2	18.2

Data compiled from reviews and research work by (Sohrab Haghighi Mood, 2013, Vishnu Menon, 2012, Alya Limayema, 2012).

Substrate from Agricultural Processing

These wastes material are generated from plants and parts of plants during production processes as by-products and also during packaging. The food industries that could benefit from anaerobic treatment include fruit and vegetable canning, edible oil refining, dairy production, seafood processing, meat processing, starch and sugar production, brewing and fermentation, juice and wine production, and wastes from biodiesel production. The sector of food industry also generates large amounts of organic waste that is characterized by a very rich nutrient composition and therefore has great potential.

2.2. Forest Biomass

Woody biomass refers to all trees and parts of trees that are availed in the production of anything whose primary raw material is wood. There are many potential sources of biodegradable feedstock from woody biomass such as wood processing and production of panels and furniture, paper and pulp processing, cardboards, branches, prunnings, chips, and leaves that fall during the changing climates among others.

There is an estimated 3870×10^6 ha of forest worldwide. Out of the global production of wood that is felled for different uses 55% is used directly as fuel, e.g., as split firewood, the remaining 45% is used as industrial raw material, but about 40% of this is used as primary or secondary process residues, suitable only for energy production. The total sustainable worldwide biomass energy potential is about 30% of total global energy consumption today (Parikka, 2004). The sixth National Forest Resources General Investigation showed that 1.75×10^8 m^2 of forest and 1.25×10^{10} m^3 of growing stock are in existence in China, from which $2.8 * 10^8$ t of biomass resources can be supplied during the production process (Zhang Peidong, 2009).

A study by (Xumeng Ge, 2014) tested the potential of using Albizia tree leaves and chips and found that the amount of volatile solids available is relatively the same with 93.7% and 98.3% respectively. This is an indication that the potential for using leaves for biogas production is viable. Specific gas yields amounting to about 290 and 190 L kg^{-1} dry solids were obtained for fresh and dry leaf biomass feedstocks, respectively, with a high methane content in the methanogenic reactors (> 70%) (H.N. Chanakya, 1993).

For specific estimation of the crop, livestock, and woody biomass residue particular to location refer to FAO estimation manual and tools (Renato Cumani, 2014b, Renato Cumani, 2014a).

2.3. Municipal Solid Waste (Organic Fraction)

Municipal waste refers to garbage collected by urban and rural councils from homesteads and industries. Municipal solid wastes are perhaps the most variable feedstock as the methane yield value depends not only on the sorting method, but also on the location and development status from which the material was sourced, the time of year of collection, size of the collection area, and the waste collection efficiency (Alastair J. Ward, 2008, Yang Yanli, 2010). Due to industrialization and the development trend in most developing countries the municipal solid waste production rates are on the rise more so in the urban areas.

The primary sources of biomass are the organic fraction of municipal solid waste from residential, commercial and industrial sources. About 30% of municipal solid waste collected is deemed suitable for anaerobic digestion as the putrescible/organic element in the municipal solid waste. According to (CEPA, 2012) two examples of this feedstock include

i.   food wastes from homes, businesses and food processing companies,
ii.   green waste such as grass, yard clippings, leaves

Table 7. Physical Characteristics of constituents of Forest woody biomass, MSW and Sludge

Substrate	TS %	VS % (% of TS)
Sludge	14.58	89.37
Fruit/vegetable waste	9.15	92.28
Food waste	19.71	82.96
Grass	92.4	85.1
Filter paper	94.8	99.8
Whole wheat bread	92.1	96.4
Leaves	93.4	84.7
Bark	89.5	95.6
Sudan Grass	27	91
Orange peelings	90.2	96.9

Data compiled from reviews and research work by (Adrie Veeken, 1999, Dong Renjie, 2013)

**Table 8. Biomethane potential and biogas yields of various constituents
of MSW and Sludge**

Substrate	Biogas Potential (lg^{-1} /Kg VS)	Methane yield (m^3 /kg VS)
Municipal Solid waste	0.20 - 0.22	
Cooked meat		0.482
Cellulose		0.356
Boiled rice		0.294
Cabbage		0.277
Mixed food waste		0.472
Mechanically sorted (fresh)		0.222
Mechanically sorted (dried)		0.215
Hand sorted		0.205
Yard waste (grass)		0.209
Yard waste (leaves)		0.123
Yard waste (branches)		0.134
Yard waste (blend)		0.143
Paper (office)		0.369
Paper (corrugated)		0.278
Paper (printed newspaper)		0.100
MSW and corn silage		0.11
MSW and cattle manure		0.03
MSW digested sludge		0.29
Primary sludge		0.590
Sudan Grass		70**

Data compiled from reviews and research work by (David P. Chynoweth, 2001, Alastair J. Ward, 2008,
      Dong Renjie, 2013)
** units expressed in Nm^3/t FM.

2.4. Sludge

Sewage sludge, the byproduct of biological wastewater treatment processes, is expected
to increase continuously in the next decade, due to increasing population connected to sewage
networks, building new waste water treatment plants, and upgrading of existing plants to meet
the more stringent local effluent regulations (Xiaohu Dai, 2013). Dewatering of sludge is vital
for transportation cost reduction and ease of centralized manipulation as high solid waste
(Xiaohu Dai, 2013).

Table 9. Carbohydrate composition of selected Forest woody biomass, MSW, and Sludge

Feedstock	Carbohydrate composition (%dry weight)		
	Cellulose	Hemicellulose	Lignin
Bamboo	49 - 50	18 - 20	23
Banana waste	13	15	14
Douglas fir	35 - 48	20 - 22	15 - 21
Eucalyptus	45 - 51	11 - 18	29
Hardwood stems	40 - 55	24 -40	18 - 25
Olive tree pruning	25.0	15.9	18.2
Grasses	25 - 40	25 - 50	10 - 30
Newspaper	40 - 55	24 - 39	18 - 30
Pine	42 - 49	13 - 25	23 - 229
Poplar wood	45 - 51	25 - 28	10 - 21
Olive tree biomass	25 .2	15. 8	19.1
Jute fibres	45 - 53	18 - 21	21 - 26
Switch grass	35 - 40	25 - 30	15 - 20
Grasses	25 - 40	25 - 50	10 - 30
Softwood stem	45 - 50	24 - 40	18 - 25
Hardwood	25 – 40	45 – 47	20 – 25
Softwood	25 – 29	40 – 45	30 – 60
Grasses	35 – 50	25 – 40	n.s.
Waste papers from chemical pulps	12 – 20	50 – 70	6 – 10
Newspapers	25 – 40	40 – 55	18 - 30
Switch grass	30 - 35	40 - 45	12
Poplar	43.8	14.8	29.1
Spruce	43.8	20.8	28.3
Oak	45.2	24.5	21.0

Data compiled from reviews and research work by (Vishnu Menon, 2012, Sohrab Haghighi Mood, 2013, Alya Limayema, 2012).

3. GENERAL ASPECTS OF BIOGAS PRODUCTION FROM HIGH-SOLID ORGANIC BIOWASTES

There are several important aspects for the development of any bioprocess in high-solid anaerobic digestion (HAD), which include selection of microorganism (mesophilic- or thermophilic-) and substrate and optimization of process parameters. Going by theoretical classification based on water activity, although only fungi and yeast were termed as suitable microorganisms for solid-state fermentation (SSF), experience has shown that bacterial cultures can be well managed and manipulated for SSF process (Selvakumar P., 1999). It is generally claimed that product yields are mostly higher in SSF as compared to submerged fermentation (A., 2003).

Distinguished from aerobic fermentation, in a high-solid anaerobic digestion, the organic polymers cannot be utilized directly by microorganisms; they are first hydrolyzed into soluble monomers or dimers by extracellular enzymes. The soluble monomers or dimers are fermented to form volatile fatty acid (VFA). Finally, the VFAs are consumed by methanogens, accompanied with methane formation. Another pathway of methane formation is the reduction of carbon dioxide.

Bioreactor configuration is another key aspect of high-solid anaerobic digestion. In high-solid anaerobic digestion, solid material is non-soluble and acts both as physical support and source of nutrients. For a given substrate such as agriculture residues, due to the existence of microorganisms' spectrum in seeding inoculums, two major issues must be taken into account; one is to achieve suitable value-addition product and/or disposal, the second could be related with the pathway of a specific product coupled to the bioreactor configuration. In the latter case, it would be necessary to select the various bioreactor configurations and the most suitable one to obtain the favorable results.

Other relevant aspects here could be the selection of process parameters and their optimization. These include physicochemical and biochemical parameters such as particle size, initial moisture, pH, pretreatment of the substrate, relative humidity, temperature of incubation, agitation, age and size of the inoculums (hydraulic/sludge retention time), supplementation of nutrients such as N, P and trace elements. They are discussed in detail in section 4.

4. BIOREACTORS SYSTEM

High solids biowaste digesters are divided into ''wet'' or ''dry'' types. Wet reactors are those with a total solids value of 16% or less, whilst dry reactors have between 22% and 40% total solids, and those that fall between wet and dry are considered semi-dry (Nikita Naik, 2013, J., 2002). The dry reactor technology is mainly used with municipal solid waste or vegetable wastes rather than with manures.

These systems can be said to fall into three main groups;

1. Batch reactors are the most simple. These are simply filled with the feedstock and left for a period that can be considered to be the hydraulic retention time, after which they are emptied. The biomass input can range from 30-40% total solids and is digested in a gas-tight container. Finished digestate is used to inoculate the dry stackable waste. Advantages include simplicity of the reactor, low maintenance requirement, and minimal capital cost. These systems are generally suitable for lignocellulosic biomass feedstock (Nikita Naik, 2013).

2. One-stage continuously fed systems: In this system all the biochemical reactions take place in one reactor. The organic loading rate of single-stage digesters depends on the ability of methanogenic bacteria to tolerate the decline in pH that results from acid production during the hydrolysis step. Continuous processes reactors require waste input in regular intervals and an equivalent removal rate of similar output. This reactor type employs relatively simple technical equipment, has a high risk of

process instability, requires at minimum 20% solid material in the tank, and reduces the substrate degradation rate (Rilling, undated, P. Vandevivere, undated).

3.   Two-stage (multi-stage) continuously fed systems where the hydrolysis/ acidification and acetogenesis/ methanogenesis processes are separated. Two-stage digesters separate the initial hydrolysis and acid-producing fermentation steps from methanogenesis allowing for higher loading rates, however, requiring additional reactors and handling systems (P. Vandevivere, undated). The solids are hydrolyzed in the first-stage reactor by re-circulation of liquid over a solid bed of crop materials. The liquid/leachate is then pumped to the second-stage reactor for further degradation. Though the technical equipment are very complex the process control is optimal with minimum risk of process instability, the retention time is also reduced and a high substrate degradation rate is observed (Rilling, undated, Koppar A., 2008).

The process configuration is presented in Figure 1. There was a comparison between single-stage UASS reactor and two-stage UASS system with an anaerobic filter. It indicated the two-stage process to be slightly more effective than the single-stage, the two-stage reactor has also been used to treat horse manure (Janina Böske, 2014).

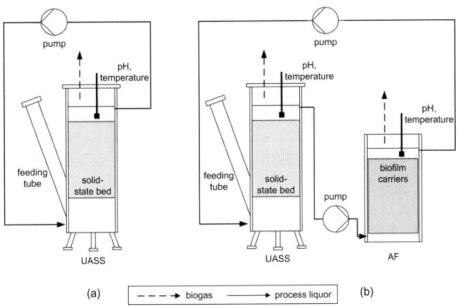

Scheme of the single-stage (a) and two-stage reactor system (b) (UASS: upflow anaerobic solid-state reactor; AF: anaerobic filter).

Figure 1. Single stage and Multi stage reactors.

4.1. Technical Advances

Different systems have diverse variables within each and the parameters associated act as operational constraints on all anaerobic digestion systems, therefore it is not possible to fully compare different systems on the exact same terms. The pretreatment options selected for different wastes and the nature of the wastes used in these systems further complicates this comparison. Biomass conversion efficiency and the biogas generation and quality can be used as an indicator for application of the system with similar feedstock. In essence systems can only be realistically designed and optimized around either waste reduction or methane production (Evans, 2001).

Continuously Stirred Tank Reactor

The CSTR is a closed-tank digester equipped with mechanical agitator for mixing of the increases surface contact between the microbes and the biomass. It is the most commonly known and used bioreactor system operated at a low total solids (TS) content, typically 5-10% TS (Banks C.J., 2011).

CSTR is already used in full-scale anaerobic digestion of high-solid biowastes. The feeding and withdrawal of reactants are conducted continuously. In this system the feedstock and microorganisms can achieve complete mixing. Crop straw, rice hull, and livestock manure are usually used as feedstock for this bioreactor. The bulky or particulate feedstocks are pretreated and/ or premixed mechanically prior to loading for the bio-reaction process. Hydrolysis and acidogenesis of biowastes are performed separately in the anaerobic digestion of high-solid biowastes such as livestock manure, food wastes and sewage sludge.

Findings of a comparison experiment of a single UASB reactor against a combination of CSTR - sedimentation tank - UASB reactor and CSTR - sedimentation tank - UASB reactor with biomass recirculation (Vasileios I. Diamantis, 2014) show the combination that has biomass recirculation to be beneficial for both increased biogas yield and drastic reduction in COD. Total nitrogen removal efficiency in the absence of organic carbon is evaluated by (Jeong-Hoon Shina, 2008) using a hollow fiber membrane biofilm reactor by autotrophic bacteria in a continuous stirred tank reactor. The SRTs and HRTs are the same in essence increasing SRT to promote higher methane yields also require longer HRT which has an impact on cost(Vipul J. Srivastava, 1988).

Upward Sludge Retention

This is a sludge retention type of digester suitable for the digestion of waste streams with high solid concentration and use plug-flow mechanisms instead of stirring. Raw materials flow into the reactor from the bottom and pass through the sludge bed. There is prolonged retention of microorganisms and unreacted solids through passive settling. The methane production rate in the USR can be increased by increasing the feed loading rate which leads to longer SRTs but shorter HRTs (Vipul J. Srivastava, 1988). A novel UASR process was developed by Mumme et al., in 2010 (J. Mumme, 2010).

Recent research on USR evaluated its performance in different temperatures (Janina Böske, 2015) and bedding material (Janina Böske, 2014). The methane production of livestock manure (horse manure) using a USR significantly upgraded by changing from mesophilic to thermophilic conditions (Janina Böske, 2015). Straw mixed with hay horse dung had the highest potential methane yields from varying mixtures of dung (hay and silage feed) and bedding material (wheat straw, flax, hemp, wood chips) (Janina Böske, 2014).

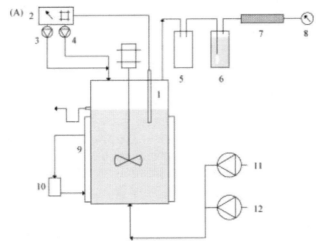

1: pH elctrode, 2: pH controller, 3: NaOH, 4: HCl: safety bottle, 6: NaOH, 7: soda lime pellets, 8: gas meter, 9: rubber hose/double –wall, 10: waterbath, 11: nutrients, 12: sucrose and sulfate at 4°C. Ref: (S.I.C. Lopes, 2008).

Figure 2. Schematic representation of the CSTR.

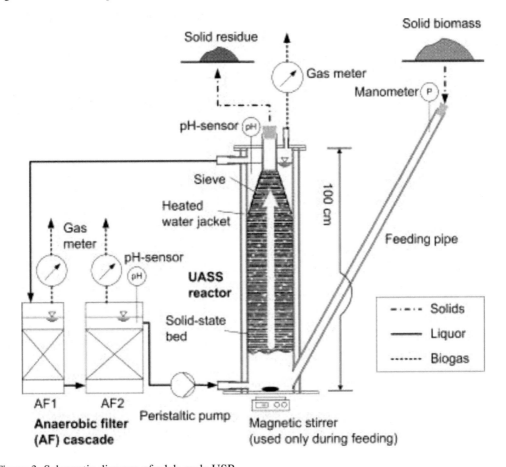

Figure 3. Schematic diagram of a lab-scale USR.

Leaching Bed Reactors (LBR)

This system was invented by David P. Chynoweth and Robert LeGand. It consists of constructed media (main body), leachate and sludge collector, and distributor of clean water or recycled leachate (Figure 3). The top section incorporates a leachate sprinkling head and headspace, the middle section is the effective reactor volume (incorporating biomass retaining vessel, sieves and meshes) and the bottom section facilitates leachate percolation and collection.

LBR sometimes called percolating anaerobic digesters are single-stage column reactors operated in batch mode. Leachate is sprayed on top of the waste and then is collected at the bottom. Leachate collected from the bottom of the reactor is circulated back into the solid waste bed to increase the moisture content, promote mass transportation, redistribute the enzymes and microbes, remove inhibitors, and minimise nutrient deficiency.

The ALBR can be loaded with high solids that have high methane yields, and low water and energy requirement. Low hydrolysis rate and prolonged start-up time and volatile fatty acid accumulation impact its overall efficiency. To overcome these limitations, a cascade process of ALBR and upflow anaerobic sludge blanket (UASB) was developed by James D. Browne (James D. Browne, 2013). A high rate of organic matter decay was achieved, but the methane yield decreased. Co-digestion of food wastes and oil palm lignocellulosic residue with pig manure was conducted using ALBR (Kanyarat Saritpongteerakaa, 2014)which enhanced biogas production.

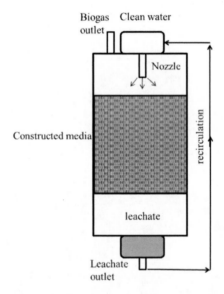

Figure 4. Schematic of Anaerobic Leach-Bed Reactor (ALBR).

Research (Craig Frear, 2014) sought to increase the efficiency of LBR and reduce the need for fresh water input to the system by adding a UASB reactor component; results had indications of ammonia stripping and leachate conditioning for use in the next LBR. Different water addition and leachate recycling strategies had varied implications on COD and water demand (Matthew N. Uke, 2013). A micro-aeration and leachate replacement method (Padma Shanthi Jagadabhi, 2010) suggest a good effect on hydrolysis rate during the digestion process. A novel two-phase continuously fed leach bed reactor (LBR) connected to an anaerobic filter (AF) with varied feeding patterns had up to 50–60% methane production increase per day, compared to constant feeding each day (Bernd Linke, 2015).

In recent cases, multi-beds were installed to ALBR with focus on the development of material and construction media (Qiyong Xua, 2014), integrating aeration to anaerobic process (Mali Sandip T., 2012), while a two-stage anaerobic digestion with leach-bed reactor(LBR) and upflow anaerobic sludge blanket reactor (UASB) was used to treat tomato, cucumber, common reed, and grass silage resulted in increased yields (Padma Shanthi Jagadabhi, 2011).

Batch Stirred Tank Reactors (BSTR)

Batch stirred tank reactors are reactors where reactants (and possibly a catalyst) are added to the reactor vessel and mixed by an agitator for a predetermined reaction time period (Australia, 2012).

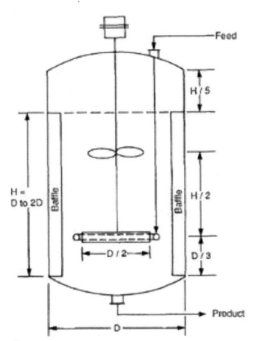

(Artin Hatzikioseyian, undated).

Figure 5. Typical proportions of a stirred batch reactor.

Continuous Feed Plug Flow Reactors (CPFR)

Tubular reactors filled with static mixing elements are often used as plug flow reactors. The baffles aid with radial mixing and cause the reacting material to take a longer path through the reactor. This allows more time for the reactants to be converted into products. In this vessel flow is continuous, usually at steady state, and configured so that conversion of the chemicals and other dependent variables are functions of position within the reactor rather than of time. In the ideal tubular reactor, the fluids flow as if they were solid plugs or pistons, and reaction time is the same for all flowing material at any given tube cross sections (Artin Hatzikioseyian, undated). Oscillatory baffled reactors (OBRs) are a form of plug flow reactor, ideal for performing long reactions in continuous mode. Three meso-reactor baffle designs (integral baffles, helical baffles and axial circular baffles were developed and computer simulations made. It was found that an increase in the net flow Reynolds number increased the optimum range of oscillatory Reynolds numbers over which plug flow can be achieved and also affected the amount of reagents required (Anh N. Phan, 2010).

Rotational Drum Fermentation System

The rotational drum fermentation system (RDFS) was developed by Jiang et al., (Wei Zhong Jiang, 2005, W.Z. Jiang, 2002). It consists of drum reactor, motor, rotational device, and stirring media. During fermentation the cylinder is rotated to bring out some form of agitation

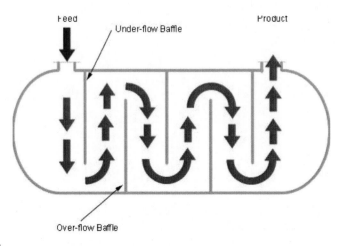

(Australia, 2012).

Figure 6. Flow of materials in a Plug Flow Reactor .

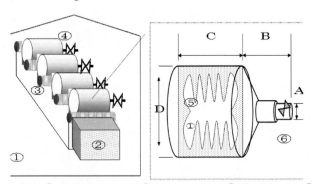

① Incubator ② DC-motor ③ Rotation device ④ Drum reactor ⑤ Baffle plate ⑥ Inlet and Outlet (A: 29mm B: 125mm C: 250mm D: 146mm)

Figure 7. Schematic diagram of a lab-scale RDFS.

During the overall anaerobic digestion of high-solid biowastes hydrolysis of particulate is a rate-limited step. Lab-scale results indicated that RDFS is favorable for quickly enhancing hydrolysis and acidogenesis of high-solid biowastes (dairy manure and food wastes). The recirculation of methanogenic leachate considerably improved acidogenic performance of solid-state biowastes (Ling Chen, 2007.). However, VFA was prone to accumulation around the particulate substrate which acted as an inhibition for further biodegradation of substrates. Water flushing (Jing Gan, 2008), ultrasonic assistant (Ling Chen, 2008) and adsorption by clay particles (Li Dawei, 2009) were integrated to the RDFS, which enhanced both the hydrolysis and acidification processes of high-solid biowastes.

Methanogenic leachate recirculation and clay particle addition were simultaneously integrated to RDFS in anaerobic digestion of food wastes (Jingwen Lu, 2013). The lab-scale results indicated that the biochemical pretreatment was predominant over the mechanical one with the more clay particle addition. On the contrary, the mechanical pretreatment was predominant with the leachate recirculation rate increasing.

Anaerobic digestion of food wastes was enhanced by removing VFA inhibition using clay minerals (Li Dawei, 2009). A particle MD reduction rate of 33.05 μm/d and a VS degradation rate of 3.53 g/L d^{-1} were obtained under an HRT ranging from 4 to 16 d (Li Dawei, 2009). For a given Y_{VA}, the addition of wheat–rice–stone (WRS) shortened the reaction time and reduced the reactor volume. The activity of microorganisms was enhanced by H$^+$ adsorption and dissociated cations in the hydrolytic and acidogenic process of solid food wastes (Cheng Fan, 2010) while enzyme production and activity was evaluated through air flow rate (A.B. Díaz, undated).

Anaerobic Fluidized Bed Reactor (AFBR)

The AFBR was found to be more stable and effective than the upflow anaerobic sludge blanket reactor (UASB) (Mehran Andalib, 2012)when dealing with high-suspended solids feedstock, this system also performed very well when used to treat municipal wastewater sludge with a high VSS destruction efficiency and COD removal efficiency (Nizar Mustaf, 2014).

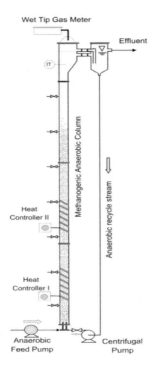

Figure 8. Schematic of AFBR.

In situ product removal (ISPR) process has been used to enhance anaerobic digestion of high-solid biowastes. A high rate anaerobic fluidized bed bioreactor (AFBR) with zeolite addition was used to treat thin stillage (Mehran Andalib, 2012) proved excellent in application. The feed solution was pumped into the bottom of the anaerobic column by a peristaltic pump, to ensure fluidization in the anaerobic column recirculation flows from top to bottom. The recirculation flows were maintained using a centrifugal pump and monitored by rate-meters; the reactor is presented in Figure 8.

4.2. Process Configuration

The shape of the reactor and material used are usually dependent on suitability and availability of technology and materials. The design is influenced by the digester parameters which are paramount to ensure growth of microorganisms by providing a conducive environment. Methanogenes are active under anaerobic conditions. The ORP of non-methanogenesis is in a range from -100 to 100 mV, while that of methanogenesis is in a range from -400 to -150 mV. Others parameters to be controlled include:

Temperature

The mean temperature and the temperature variation range (highest vs. lowest daily temperatures) have the greatest influence. The anaerobic digestion process can happen at different temperatures, which are divided into (Jian Lin Chen, 2014):

a. psychrophilic (below 20°C),
b. mesophilic (25°C – 45°C),
c. thermophilic (45°C –65°C)

Most anaerobic reactors operate at either mesophilic or thermophilic temperatures, with optimum temperatures at 37 °C and 55 °C degrees respectively, because the biomass activities and anaerobic treatment capacities are significantly reduced under psychrophilic conditions. Generally, the rate of chemical reaction increases along with the increase of the surrounding temperature.

Mesophilic bacteria are able to tolerate ±3°C differences without significant reductions of the methane production, thermophilic bacteria are more sensitive to the temperature fluctuation where ±1°C differences affected the methane production negatively (LiShan, 2014, J., 2002).

Operation of a digester in the mesophilic range is more stable, and consumes less energy and experiences less ammonia inhibition. However mesophilic microorganisms are slower resulting in need for a longer retention time in the digester. Thermophilic digestion is regarded as the more efficient method as it offers an enhanced hydrolysis process and shorter retention time with a higher biogas yield. The increase in methane yield from the thermophilic process has to be balanced against the increased energy requirement for maintaining the reactor at the higher temperature (World, 2014).

pH

Literature suggests the optimum pH for a stable anaerobic digestion process and high biogas to be in the range of 6.5-7.5(Alastair J. Ward, 2008). An alkalinity level of approximately 3000 mg /L has to be available at all times to maintain sufficient buffering capacity (Yvonne Vögeli, 2014). Although the optimal pH of methanogenesis is around pH 7.0, the optimum pH of hydrolysis and acidogenesis has been reported as being between pH 5.5 and 6.5 hence the introduction of two stage systems. Buffer capacity is often referred to as alkalinity in anaerobic digestion, which is the equilibrium of carbon dioxide and bicarbonate ions that provides resistance to significant and rapid changes in pH and is a more reliable method of measuring digester imbalance than direct measurements of pH (Alastair J. Ward, 2008).

Pretreatment

Pretreatment increases accessible surface area, decrystalizes cellulose, removes hemicelluloses, and alters the lignocellulosic structure of biowastes; which enhances the action of enzymes or microorganism.

There are several ways of feedstock pretreatment with varying application to feedstock as well as impact on the biogas methane production. Figure 9 shows the effect of particle size on the methane yield of sisal fiber wastes (A. Mshandete, 2006). In CSTR systems for example some of the common pretreatment options available include Physical (pulverization, stream-explosion, pyrohydrolysis, extrusion, and irradiation), chemical (acidic, alkali, and ionic liquid) and biological (use of fungi and enzymes) pretreatment; application of some of these methods is as indicated in Table 10 (Liangcheng Yang, 2015).

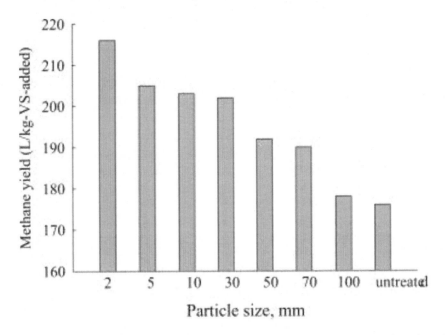

Figure 9. Methane yields from AD of sisal fiber waste with different particle sizes (Liangcheng Yang, 2015).

Table 10. Pretreatment methods of biowastes for biogas enhancement

Pretreatment	Feedstock	CH4yield (L/kg VS)	Increase (%)
Steam explosion	Rape straw	257	17
	Corn stalks/straw	220-484	17-63
	Seaweed	268	20
	Bulrush	205	25
	Wheat straw	331	30
	Hardwood	234-369	46-80
Hot water	Greenhouse residues	274	19
	Wheat straw	94	20
	Oil palm empty fruit bunches	208	40
	Rice straw	133	222
Dilute acid	Sunflower stalks	233-248	21-29
	Sunflower oil cake	289	48
	Herbal-extraction residue	N/A	100
	Sugarcane bagasse	200	166
Wet oxidation	Winter rye straw	482	34
	Miscanthus/ willow	360	80
	Yard waste	685	99
Alkaline	Rice straw	341-520	30-100
	Corn stover	211-372	37-73
	Wheat straw	166-289	39-112
	Hardwood	360	57
	Switchgrass	490	65
	Pine tree wastes	75-107	184-224
	Softwood	210	600
Biological	Sweet chestnut leaves/hay	N/A	15
	Cotton stalks	N/A	25
	Cassava residues	259	97
	Sisal leaf decortications residue	292	101
	Yard trimmings	45	154
	Japanese cedar wood chips	43	400

Data from review work by (Liangcheng Yang, 2015).

Carbon to Nitrogen Ratio

The C:N ratio is an important parameter in estimating nutrient deficiency and ammonia inhibition. Optimal C:N ratios in anaerobic digesters are between 16 and 30 (Tong Zhang, 2013, Alastair J. Ward, 2008). A high C:N ratio is an indication of rapid consumption of nitrogen by methanogens, whereas, a low C:N ratio causes ammonia accumulation and pH values then may exceed 8.5 which may be toxic.

Co-digestion can be applied to balance the C:N ratio while improving nutrient balance, decreasing the effect of toxic compounds on the digestion process, and conditioning the rheological qualities of the substrate. It also enhances the methane yield (World, 2014).

CSTR was adapted to the anaerobic co-digestion of chicken manure (CM) and corn stover (CS) (Yeqing Li, 2014) which enhanced the production of biogas and energy recovery efficiency. It was also established that integrated utilization of digestate could increase energy recovery efficiency up to 85%. Food waste for example has TS content between 20% and 30%. Treating it with CSTR requires dilution with water or agricultural slurry to facilitate homogenization and mixing. This demands a large amount of energy due to the relatively large heat capacity of water (Jagadabhi P.S., 2011). The co-digestion of livestock manure and food waste can overcome this problem.

In an experiment of co-digestion with goat manure and three different types of crop residue mixtures to achieve various C: N ratios proved an effective way to prolong the period of the highest gas production while improving biogas yield. The results showed goat manure/ wheat straw co-digestion had an increase of 22% up to 62% compared with single wheat straw and goat manure. The same trends were observed for the goat manure/corn straw and goat manure/ricestraw treatments, which had considerably higher increases (Tong Zhang, 2013).

Inoculation and Start-Up

Bacteria necessary for anaerobic digestion has to be introduced at the beginning; this can be done by using slurry from a well-functioning plant or by use of prepared inoculum. Acclimatization of bacteria to feedstock can be achieved by progressively increasing the daily feeding load which allows time to achieve a balanced microorganism population. Overloading results from either feeding too much biodegradable organic matter compared to the active population capable of digesting it, or rapidly altering digester conditions (e.g., abrupt change of temperature, accumulation of toxic substances, flow rate increase). Such disturbances specifically affect methanogenic bacteria, whereas the acidogenic bacteria, which are more tolerant, continue to work, and produce acids. This eventually leads to an acidification of the digester which inhibits the activity of methanogens. Addition of manure can avoid this as it increases the buffer capacity, thereby reducing the risk of acidification (Yvonne Vögeli, 2014).

A quick-start of anaerobic digestion may help the process be more stable and efficient; this may include enlarging the ratio of seeding to substrate, diluting the substrate by adding water or recirculating effluent of sound reactor. Anders (Anders Lagerkvist, 2015) proposed a novel procedure for quick-start of biowastes' AD using aeration. A conventional 1300 m^3 CSTR was fed sewage sludge and food waste. First the thick and slightly acidic sewage was fed daily to the reactor for 8 days then compressed air was injected into reactor for a month this reduced the methanogenesis start up by a week.

Organic Loading Rate

The Organic Loading Rate (OLR) is a measure of the biological conversion capacity of the anaerobic digestion system. It represents the substrate quantity introduced into the reactor volume in a given time. OLR is a particularly important control parameter in continuous systems, as overloading leads to a significant rise in volatile fatty acids which can result in acidification and system failure as described above.

Economically, the biogas plants are designed to get the maximum methane production within a shorter period of time, instead of getting the maximum amount of methane from complete decomposition of the organic constituents which may need a very long retention time in the digester. Organic loading rate also indicates how many kilograms of organic dry matter are fed into the digester per cubic meter of working volume per unit of time.

Hydraulic Retention Time

The Hydraulic Retention Time quantifies the time the liquid fraction remains in the reactor. It is calculated by the ratio of the reactor (active slurry) volume to the input flow rate of feedstock. The hydraulic retention time required to allow complete anaerobic digestion reactions varies with different technologies, process temperature and waste composition. Recommended hydraulic retention time for wastes treated in a mesophilic digester range from 10 to 40 days. Lower retention times down to a few days only, are required in digesters operated in the thermophilic range (Verma, 2002). A distinction should made between Hydraulic Retention Time and Solids Retention Time, but for digestion of solid waste, HRT and SRT are generally considered equal (Yvonne Vögeli, 2014).

Mixing

Mixing is mainly done so as to:

1. Blend the fresh material with digestate or inoculums
2. Prevent scum formation in the digester: a very thick impermeable scum may prevent gas release leading to structure failure
3. Maintain a nearly homogeneous temperature zone in the digester substrate

Mixing and stirring equipment, and the way it is performed, varies according to reactor type and TS content in the digester (Yvonne Vögeli, 2014). Various alternate mixing methods such as mechanical agitation, recirculation of biogas through the bottom of the reactor or hydraulic mixing by recirculation of the fermented liquid with a pump can be used to distribute substrates, enzymes and microorganisms throughout the digester(World, 2014, Rilling, undated). The contents of most anaerobic digesters are mixed to ensure efficient transfer of organic material for the active microbial biomass, to release gas bubbles trapped, and to prevent sedimentation of denser particulate material. Mixing does not always take place continuously.

Inhibition

An inhibitor causes an adverse shift in the microbial population or may inhibit growth off bacteria (Ye Chen, 2008). Some compounds at high concentrations can be toxic to the anaerobic process. Inhibition depends on the concentration of the inhibitors, the composition of the substrate and the adaptation of the bacteria to the inhibitor (Deublein, 2008). Co-digestion with other wastes, adaptation of microorganisms to inhibitory substances, and incorporation of methods to remove or counteract toxicants before anaerobic digestion can significantly improve waste treatment efficiency.

The following is an overview of the common anaerobic digestion inhibitors;

Chlorophenols include monochlorophenols (CPs), dichlorophenols (DCPs), trichlorophenols (TCP), tetrachlorophenols (TeCPs), and pentachlorophenol (PCP). Chlorophenols are used widely as pesticides, herbicides, antiseptics and fungicides as well as preservatives for wood, glue, paint, vegetable fibers, and leather (Jian Lin Chen, 2014, Ye

Chen, 2008). Toxicity of chlorophenols depends on their degree of chlorination, the position of the chlorine, and the purity of the sample with inhibition increasing proportional to the number of chlorine substitutions. Biological dechlorination is the main process for PCP removal in anaerobic digestion (Jian Lin Chen, 2014). Chlorophenols are toxic to many organisms by disruptingthe proton gradient across membranes and interferingwith energy transduction of cells (Ye Chen, 2008).

Halogenated aliphatics (HAs) are organic chemicals in which one or more hydrogen atoms have been replaced by a halogen. HAs are used in industry as solvents, chemical intermediates, and fumigants and insecticides. Hey exist as polychlorinated aliphatic hydrocarbons, such as dichloromethane (DCM, CH_2Cl_2), chloroform (CF, $CHCl_3$), trichloroethylene (TCE, C_2HCl_3), and perchlorethylene (PCE, C_2Cl_4), (Jian Lin Chen, 2014). Halogenated aliphatics are removed by microorganisms that utilize dehalogenase enzymes.

Long chain fatty acids (LCFAs) are fats that have several carbons in their chain and include unsaturated and saturated fats. LCFAs are mostly found in wastes from food processing industries because they are rich in fats, proteins, and carbohydrates. This may include wastes such as oil/fat wastewater, ice-cream wastes , dairy wastewater, fish wastes, slaughterhouse wastewater, and vegetable waste. Inhibition by LCFAs of anaerobic processes depends on the type of LCFA, the microbial population, and the temperature. They cause inhibition on methanogenic bacteria and create sludge over the surface of the substrate. Creating competition using synthetic adsorbents has been indicated as a way to reduce inhibition (Jian Lin Chen, 2014).

Ammonia is vital for the growth of microorganisms involved in anaerobic digestion but in excess amounts it becomes inhibitory. The fermentation of nitrogen-containing materials such as urea and proteins like most animal wastes releases ammonia-nitrogen. Amounts of ammonia inhibition levels can be associated to the differences in substrates and inocula, environmental conditions (temperature, pH), and acclimation periods. Free ammonia causes proton imbalance, and/or potassium (K^+) deficiency, while ionized ammonium usually inhibits the methane synthesizing enzyme directly; this form of inhibition also raises the maintenance energy. A common approach to ammonia inhibition relies on dilution, increasing the biomass retention in the reactor, pH adjustment, air stripping, immobilization of microorganisms, addition of ionic adsorbents (Jian Lin Chen, 2014).

Sulfide is present in industrial waste waters. Toxicity is through inhibiting sulfate reducing bacteria and methane producing bacteria, reducing the rate of methangenesis, and decreasing the quantity of methane produced by competing for the available carbon and/or hydrogen. Sulfide toxicity to methanogens is proportional to its concentration in the substrate and H_2S concentration in the gas phase (Jian Lin Chen, 2014). Therefore, diluting the wastewater stream can prevent toxicity, incorporating a sulfide removal step such as stripping, coagulation, oxidation, precipitation, and biological conversion to elemental sulfur. There is a relationship between ammonia inhibition and sulfide inhibition.

Heavy metals are often present in industrial wastewaters and municipal sludge. These include copper (Cu), zinc (Zn), lead (Pb), mercury (Hg), chromium (Cr), cadmium (Cd), iron (Fe), nickel (Ni), cobalt (Co) and molybdenum (Mo). These metals are non-biodegradable. The potential toxicity of heavy metals is determined by the total metal concentration, chemical forms of the metals, and process-related factors such as pH and redox potential, It is believed that heavy metals show their toxicity due to their disruption of enzyme function and structure by binding with thiol and other groups on protein molecules, or by replacing natural metals in enzyme prosthetic groups. The most important methods for mitigating heavy metal toxicity are precipitation by sulfide, sorption, and chelation by organic and inorganic ligands (Ye Chen, 2008, Jian Lin Chen, 2014).

Light metals ions (Na, K, Mg, Ca, and Al) also cause inhibition in AD. Like ammonia, salts are crucial for microorganisms but when the threshold is exceeded they are toxic. High salt levels cause bacterial cells to dehydrate due to osmotic pressure. Seafood processing wastewaters contain high concentrations of different cations and anions (Ye Chen, 2008).

Biowaste Initial Water Content

Biogas production and the water content of the initial material are interdependentdue to the fact that bacteria take up available substrates to work on in dissolved form (Rilling, undated). Crop residues of wheat, rice, and maize for example should have a dry matter content of 35% before being chopped and ensiled (Dong Renjie, 2013).

Degree of Degradation

The degree of degradation provides information about the efficiency of the substrate converted by biological and chemical degradation of organic compounds. The degradation rate is mainly based on metabolic processes. There are many ways to define the degradation rate, either as single component or as sum parameter, e.g., COD (chemical oxygen demand) and organic dry matter content (LiShan, 2014).

5. PERSPECTIVE

In general, the basic requirements of an anaerobic digester design are: to allow for a continuously high and sustainable organic load rate, a short hydraulic retention time (to minimise reactor volume), and to produce the maximum volume of methane. Reactor shape must take into consideration the construction practicalities of both mixing and heat loss (Alastair J. Ward, 2008).

Pretreatment techniques will continue to be the focus in future research and development. Enhancing hydrolysis and acidogenesis of high-solid biowastes prevails over methanogenesis by adjusting reactor and/or process configurations. It is a promising option for stimulating both hydrolysis and methanogenesis using novel additives such as mineral clay, coal ash, charcoal, and grass peat especially in rural areas. Water consumption should be considered in process optimization design, as well.

REFERENCES

A. Mshandete, L. B., Kivaisi, A. K., Rubindamayugi, M. S. T. &Mattiasson,B. (2006). Effect Of Particle Size On Biogas Yield From Sisal Fibre Waste. *Renewable Energy,31*, 2385 – 2392.

P, A. (2003). Solid-State Fermentation. *Biochemical Engineering Journal,13*, 81-84.

A.B. Díaz, I. D. O., Caro,I. & Blandino Undated, A. Solid State Fermentation In A Rotating Drum Bioreactor For The Production Of Hydrolytic Enzymes.

Adrie Veeken, B. H. (1999). Effect Of Temperature On Hydrolysis Rates Of Selected Biowaste Components. *Bioresource Technology,69*, 249 - 254.

Alastair, J., Ward, P. J. H., Peter, J. Holliman& David, L. Jones.(2008). Optimisation Of The Anaerobic Digestion Of Agricultural Resources. *Bioresource Technology,99*, 7928 - 7940.

Alya Limayema, S. C. R. (2012). Lignocellulosic Biomass For Bioethanol Production: Current Perspectives, Potential Issues And Future Prospects. *Progress In Energy And Combustion Science,38*, 449 - 467.

Anders Lagerkvist, M. P. & Tommy, Wikström.(2015). Quick-Start Of Full-Scale Anaerobic Digestion (Ad) Using Aeration. *Waste Management.*

Anh, N. &Phan, A. H. (2010). Development And Evaluation Of Novel Designs Of Continuous Mesoscale Oscillatory Baffled Reactors. *Chemical Engineering Journal,159*, 212 - 219.

Artin Hatzikioseyian, E. R. Undated. *Bioreactors For Metal Bearing Wastewater Treatment* [Online]. Available: Http://Www.Metal.Ntua.Gr/~Pkousi/E-Learning/Bioreactors/Page_ 07. Htm.

Australia, C. O. (2012). *Start Up Reaction System* [Online]. Available: Https://Nationalvetcontent.Edu.Au/Alfresco/D/D/Workspace/Spacesstore/Dc6a7f1f-E3d3-44d8-9bf1-984ea8cb3c01/204/Pmaops302b/Proc302-010200-Types-Of-Reactors.Htm.

Banks C.J., C. M., Heaven, S. & Arnold R. (2011). Anaerobic Digestion Of Source-Segregated Domestic Food Waste: Performance Assessment By Mass And Energy Balance. *Bioresource Technology, 102*, 612–620.

Bartali Elhoussine, B. M., Amal, Saber Mohamed& Yasser, A. El-Tahlawy.(2014). Report On Biowaste Management For The Selected Feedstock. Biowaste4sp.

Bernd Linke, Á. R.-A., Carsten, Jost&Andreas, Krieg.(2015). Performance Of A Novel Two-Phase Continuously Fed Leach Bed Reactor For Demand-Based Biogas Production From Maize Silage. *Bioresource Technology,177*, 34 - 50.

Cheng Fan, L. M., Li Dawei, Chen Ling, Jiang Weizhong, Kitamura Yutaka& Li, Baoming.(2010). Volatile Organic Acid Adsorption And Cation Dissociation By Porphyritic Andesite For Enhancing Hydrolysis And Acidogenesis Of Solid Food Wastes. *Bioresource Technology,101.*

Craig Frear, T. E., Liang Yu, Jingwei Ma& Shulin Chen.(2014). Two Novel Floor-Scale Anaerobic Digester Systems For Processing Food Waste. Washington State University.

David, P., Chynoweth, J. M. O.& Robert, Legrand.(2001). Renewable Methane From Anaerobic Digestion Of Biomass. *Renewable Energy,22*, 1 - 8.

Deublein, D.& Steinhauser,A. (2008). Biogas From Waste And Renewable Resources. *Wiley - Vch.*

Dong Renjie, B. R. (2013). *Biogas Engineering And Application: Guide To Biogas- From Production To Use*, China Agricultural University Press.

E. Aymerich, M. E.-G.& Sancho,L. (2013). Analysis Of The Stability Of High-Solids Anaerobic Digestion Of Agro-Industrial Waste And Sewage Sludge. *Bioresource Technology,144*, 107 - 114.

Evans, G. (2001). Biowaste And Biological Waste Treatment.

FAO.(2014). Faostat.

H.N. Chanakya, S. B., Rajan, M. G. C. &Wahi,M. (1993). Two Phase Fermentation Of Whole Leaf Biomass To Biogas. *Biomass And Bioenergy.*

J., M.-A. (2002). Biomethanization Of The Organic Fraction Of The Municipal Solid Wastes. London, Uk: Iwa Publishing, .

J. Mumme, B. L.& Tölle., R. (2010). Novel Upflow Anaerobic Solid-State (UASS) Reactor. *Bioresource Technology*, *101*, 592-599.

Jagadabhi P.S., P., K. &Rintala, J. (2011). Two-Stage Anaerobic Digestion Of Tomato, Cucumber, Common Reed, And Grass Silage In Leach-Bed Reactors And Upflow Anaerobic Sludge Blanket Reactors. *Bioresource Technology*,*102*, 4726–4733.

James, D., Browne, E. A.& Jerry, D. Murphy.(2013). Improving Hydrolysis Of Food Waste In A Leach Bed Reactor. *Waste Management*, *33*, 2470–2477.

Janina Böske, B. W., Felix Garlipp, Jan Mumme& Herman, Van Den Weghe.(2015). Upflow Anaerobic Solid-State (Uass) Digestion Of Horse Manure: Thermophilic Vs. Mesophilic Performance. *Bioresource Technology*, *175*, 8-16.

Janina Böske, B. W., Felix Garlipp, Jan Mumme & Herman Van Den Weghe. (2014). Anaerobic Digestion Of Horse Dung Mixed With Different Bedding Materials In An Upflow Solid-State (Uass) Reactor At Mesophilic Conditions. *Bioresource Technology*, *158*, 111-118.

Jeong-Hoon Shina, B.-I. S., Yun-Chul Chunga& Youn-Kyoo Choung.(2008). A Novel Cstr-Type Of Hollow Fiber Membrane Biofilm Reactor For Consecutive Nitrification And Denitrification. *Desalination*,*221*, 526 - 533.

Jian Lin Chen, R. O., Terryw, J. Steele& David C. Stuckey.(2014). Toxicants Inhibiting Anaerobic Digestion: A Review. *Biotechnology Advances*,*32*, 1523 - 1534.

Jing Gan, L. C., Baoming Li, Weizhong Jiang& Yutaka Kitamura.(2008). A Rotational Drum Fermentation System With Water Flushing For Enhancing Hydrolysis And Acidification Of Solid Organic Wastes. *Bioresource Technology*,*99*, 2571 - 2577.

Jingwen, Lu D. L., Ling Chen, Yutaka Kitamura, Weizhong Jiang& Baoming Li.(2013). Simultaneous Pretreatment And Acidogenesis Of Solid Food Wastes By A Rotational Drum Fermentation System With Methanogenic Leachate Recirculation And Andesite Porphyry Addition. *Bioresource Technology*,*138*, 101 - 108.

Kanyarat Saritpongteerakaa, P. B., Shihwu Sungc& Sumate Chaiprapat. (2014). Co-Fermentation Of Oil Palm Lignocellulosic Residue With Pig Manure In Anaerobic Leach Bed Reactor For Fatty Acid Production. *Energy Conversion And Management*, *84*, 354-362.

Koppar A., P. P. (2008). Single-Stage, Batch, Leach Bed Thermophilic Anaerobic Digestion Of Spent Sugar Beet Pulp. *Bioresource Technology 99*, 2831–2839.

Li Dawei, Z. T., Chen Ling, Jiang Weizhong, Cheng Fan, Li Baoming, Kitamura Yutaka.(2009). Using Porphyritic Andesite As A New Additive For Improving Hydrolysis And Acidogenesis Of Solid Organic Wastes. *Bioresource Technology*,*100*, 5594 - 5599.

Liangcheng Yang, F.& Xumengge, Yeboli.(2015). Challenges And Strategies For Solid-State Anaerobic Digestion Of Lignocellulosic Biomass. *Renewable And Sustainable Energy Reviews*,*44*, 824 - 834.

Ling Chen, B. L., Dawei Li, Jing Gan, Weizhong Jiang& Yutaka Kitamura.(2008). Ultrasound-Assisted Hydrolysis And Acidogenesis Of Solid Organic Wastes In A Rotational Drum Fermentation System. *Bioresource Technology*,*99*, 8337 - 38343.

Ling Chen, W. Z. J., Yutaka Kitamura& Baoming Li.(2007). Enhancement Of Hydrolysis And Acidification Of Solid Organic Waste By A Rotational Drum Fermentation System With Methanogenic Leachate Recirculation. *Bioresource Technology*, *98*, 2194-2200.

Lishan, W. (2014). *Biogas Production From Presorted Biowaste And Municipal Solid Waste From Sweden Substrate Characterization, Wet Fermentation, And Cash Flows Analysis.* Bio- And Environmental Engineering, Ostfalia University Of Applied Sciences.

Mali Sandip T.,Biradar,K. C. &Ashok, H. (2012). Enhancement Of Methane Production And Bio-Stabilization Of Municipal Solid Waste In Anaerobic Bioreactor Landfill *Bioresource Technology*,*110*, 10 - 17.

Matthew N. Uke& Edward Stentiford. (2013). Enhancement Of The Anaerobic Hydrolysis And Fermentation Of Municipal Solid Waste In Leachbed Reactors By Varying Flow Direction During Water Addition And Leachate Recycle. *Waste Management*,*33*, 1425 - 1433.

Mehran Andalib, H. H., Elsayed Elbeshbishy, George Nakhla& Jesse Zhu (2012). Treatment Of Thin Stillage In A High-Rate Anaerobic Fluidized Bed Bioreactor (AFBR). *Bioresource Technology*,*121*, 411–418.

Nikita Naik, E. & Tkachenko, Roy Wung. (2013). Digestion Of Organic Municipal$Solid Waste In California.

Nizar Mustaf, E. E., George Nakhlaa, Jesse Zhu. (2014). Anaerobic Digestion Of Municipal Wastewater Sludges Using Anaerobic Fluidized Bed Bioreactor. *Bioresource Technology*, *172*, 461-466.

Vandevivere, P. & Undated, L. D. B. A. W. V. Types Of Anaerobic Digesters For Solid Wastes.

Padma Shanthi Jagadabhi, Prasad Kaparaju& Jukka Rintala. (2010). Effect Of Micro-Aeration And Leachate Replacement On Cod Solubilization And Vfa Production During Mono-Digestion Of Grass-Silage In One-Stage Leach-Bed Reactors. *Bioresource Technology*,*101*, 2818 - 2824.

Padma Shanthi Jagadabhi, P. K. & Jukka Rintala. (2011). Two-Stage Anaerobic Digestion Of Tomato, Cucumber, Common Reed, And Grass Silage In Leach-Bed Reactors And Upflow Anaerobic Sludge Blanket Reactors. *Bioresource Technology*, *102*, 4726-2733.

Parikka, M. (2004). Global Biomass Fuel Resources. *Biomass And Bioenergy*,*27*, 613 - 620.

Qiyong Xua, X. J., Zeyu Ma, Huchun Tao& Jae Hac Ko. (2014). Methane Production In Simulated Hybrid Bioreactor Landfill. *Bioresources Technology*,*168*, 92-96.

Rapport J.L., Z. R. H., Jenkins B.M.,Hartsough B.R.&Tomich T.P.(2011). Modeling The Performance Of The Anaerobic Phased Solids Digester System For Biogas Energy Production. . *Biomass Bioenergy*, *35*, 1263–1272.

Renato Cumani, A. K., Harinder Makkar, Walter Kollert, Seth Meyer, Francesco Tubiello And His Team, Alessio D'amato (University Of Rome, Tor Vergata) & Luca Tasciotti. (2014a). Bioenergy And Food Security Rapid Appraisal (BEFS RA) User Manual Forest Harvesting And Wood Processing Residues. Fao.

Renato Cumani, A. K., Harinder Makkar, Walter Kollert, Seth Meyer, Francesco Tubiello And His Team, Alessio D'amato (University Of Rome, Tor Vergata) & Luca Tasciotti. (2014b). Bioenergy And Food Security Rapid Appraisal (BEFS RA) User Manual Crop Residues And Livestock Residues. Fao.

Richard, T. (1996). The Effect Of Lignin On Biodegradability. *Cornell Composting*.

Rilling, N. Undated. Anaerobic Fermentation Of Wet And Semidry Garbage Waste Fractions.

S.I.C. Lopes, C. D., Capela, M. I. &Lens, P. N. L.(2008).   Comparison Of Cstr And Uasb Reactor Configuration For The Treatment Of Sulfate Rich Wastewaters Under Acidifying Conditions. *Enzyme And Microbial Technology*,*43*, 471 - 479.

Selvakumar P., P. A.(1999).   Solid-State Fermentation For The Synthesis Of Inulinase From The Strains Of Staphylococcus Sp. And Kluyveromycesmarxianus,. *Process Biochemistry*,*34*, 851-855.

Sohrab Haghighi Mood, A. H. G., Meisam Tabatabaei, Gholamreza Salehijouzani, Gholam Hassan Najafi, Mehdi Gholami& Mehdi Ardjmand. (2013).   Lignocellulosic Biomasstobioethanol,Acomprehensivereview With A Focus On Pretreatment. *Renewable And Sustainable Energy Reviews*,*27*, 77 - 93.

Thomas Amon, B. A., Vitaliy Kryvoruchko, Andrea Machmüller, Katharina Hopfner-Sixt, Vitomir Bodiroza, Regina Hrbek, Jürgen Friedel, Erich Pötsch, Helmut Wagentristl, Matthias Schreiner& Werner Zollitsch. (2007).   Methane Production Through Anaerobic Digestion Of Various Energy Crops Grown In Sustainable Crop Rotations. *Bioresource Technology*,*98*, 3204 - 3212.

Tong Zhang, L. L., Zilin Song, Guangxin Ren, Yongzhong Feng, Xinhui Han& Gaihe Yang. (2013).   Biogas Production By Co-Digestion Of Goat Manure With Three Crop Residues. *Plos One*, 8.

Vasileios I. Diamantis, A. G. K., Spyridon Ntougias, Vasiliki Tataki, Paraschos Melidis& Alexander Aivasidis. (2014).   Two-Stage CSTR–UASB Digestion Enables Superior And Alkali Addition-Free Cheese Whey Treatment. *Biochemical Engineering Journal*,*84*, 45 - 52.

Verma, S. (2002).   *Anaerobic Digestion Of Biodegradable Solids In Municipal Solid Wastes.* Earth Resources Engineering, Columbia University.

Vipul J. Srivastava, K. F. F., David, P. Chynoweth& James, R. Frank. (1988).   Improved Efficiency And Stable Digestion Of Biomass In Nonmixed Upflow Solids Reactors. *Institute Of Gas Technology.*

Vishnu Menon, M. R. (2012).   Trends In Bioconversion Of Lignocellulose: Biofuels, Platform Chemicals & Biorefinery Concept. *Progress In Energy And Combustion Science*,*38*, 522 - 550.

W.Z. Jiang, Y. K., Ishizuka, N. &Shiina,T. (2002).   A Rotational Drum Fermentation System For Dry Methane Fermentation (2): Effect Of Hydraulic Retention Time (HRT) And Stirring Media In Fermentor On Acidogenic Process. *Journal Of The Society Of Agricultural Structures, Japan*,*33*, 189–196.

Wei Zhong Jiang, Y. K.& Baoming Li. (2005).   Improving Acidogenic Performance In Anaerobic Degradation Of Solid Organic Waste Using A Rotational Drum Fermentation System. *Bioresource Technology*,*96*, 1537 - 1543.

Wilkie, A. C. (2013).   Biogas: A Renewable Biofuel.

World, W. M. (2014).   Biowaste: Dry Advice.

Xiaohu Dai, N. D., Bin Dong, Lingling Dai. (2013).   High-Solids Anaerobic Co-Digestion Of Sewage Sludge And Food Waste In Comparison With Mono Digestions: Stability And Performance. *Waste Management*, 33, 308 - 316.

Xumeng Ge, T. M., Lisa Keith, Yebo Li. (2014).   Biogas Energy Production From Tropical Biomass Wastes By Anaerobic Digestion. *Bioresource Technology*,*169*, 38 - 44.

Yang Yanli, Z. P., Zhang Wenlong, Tian Yongsheng, Zheng Yonghong &Wang Lisheng. (2010).   Quantitative Appraisal And Potential Analysis For Primary Biomass Resources

For Energy Utilization In China. *Renewable And Sustainable Energy Reviews,14*, 3050 - 3058.

Ye Chen, J. J. C.& Kurt S. Creamer. (2008). Inhibition Of Anaerobic Digestion Process: A Review. *Bioresource Technology,99*, 4044 - 4064.

Yeqing Li, R. Z., Yanfeng He, Chenyu Zhang, Xiaoying Liu, Chang Chen& Guangqing, Liu. (2014). Anaerobic Co-Digestion Of Chicken Manure And Corn Stover In Batch And Continuously Stirred Tank Reactor (CSTR). *Bioresource Technology, 156*, 342-347.

Yvonne Vögeli, C. R. L., Amalia Gallardo, Stefan Diener& Christian Zurbrügg. (2014). *Anaerobic Digestion Of Biowaste In Developing Countries Practical Information And Case Studies.*

Zhang Peidong, Y. Y., Tian Yongsheng, Yang Xutong, Zhang Yongkai, Zheng Yonghong& Wang Lisheng. (2009). Bioenergy Industries Development In China: Dilemma And Solution. *Renewable And Sustainable Energy Reviews,13*, 2571 - 2579.

In: Gas Biofuels from Waste Biomass
Editor: Zhidan Liu

ISBN: 978-1-63483-192-5
© 2015 Nova Science Publishers, Inc.

Chapter 6

BIOHYDROGEN PRODUCTION THROUGH DARK FERMENTATION: A REVIEW

Buchun Si and Zhidan Liu[*]

Laboratory of Environment-Enhancing Energy (E2E), College of Water Resources
and Civil Engineering, China Agricultural University, Beijing, China

ABSTRACT

Hydrogen production through dark fermentation is one of the promising approaches through biological route. Extensive work on bench and pilot scale with promising prospects have been reported. In this chapter, the publications related to the biohydrogen production by dark fermentation paper were reviewed, including the pathways, kenitic models and the key factors such as inoculum, temperature, pH, feedstock and reactors types. Based on the reviews of advances in the research, this chapter tried to conclude the optimal values of each factors for dark fermentation. Moreover, dark fermentation combines with other process such as biomethane fermentation and microbial fuel cells; these are also discussed. Finally, challenges and prospects on biohydrogen production through dark fermentation were addressed.

Keywords: biohydrogen, dark fermentation, solid waste, wastewater

1. INTRODUCTION

Hydrogen is regarded as a promising alternative to conventional fossil fuels because it has the potential to eliminate all of the problems that the fossil fuels create. Hydrogen can be used directly in internal combustion engines or used to produce electricity through fuel cells with high efficiency (Lin et al., 2012). Hydrogen has special properties as a transportation fuel, including a rapid burning speed, high effective octane number, high combustion efficiency in automobiles (50% more efficient than gasoline), and high energy content (2.75

---

[*] Email: zdliu@cau.edu.cn

times of any hydrocarbons) (Balat, 2008). In addition, hydrogen is a clean fuel which produces pure water after combustion, and has no toxicity or ozone-forming potential (Li and Fang, 2007). Hydrogen also has a wide range of industrial applications. It can be used for the syntheses of ammonia, alcohols, and aldehydes, as well as for the hydrogenation of edible oil, petroleum, coal, and shale oil.

Currently, 96% of world hydrogen production comes from fossil fuels. Among them, natural gas is the main raw material (> 75% of production) and steam reforming is the most used method. This process is far of being sustainable due to its high greenhouse gas (GHG) emissions (Dufour et al., 2009). Compare with fossil hydrogen production pathways, the biohydrogen pathways have the less energy ratios (ER) and GHS emissions. Consequently, biohydrogen is worthy of consideration in the planning and development of a hydrogen economy, both from an energy and from an environmental perspective (Djomo and Blumberga, 2011). Currently, known biohydrogen production processes are classified as direct biophotolysis, photo-fermentations, indirect biophotolysis and dark fermentation (Argun and Kargi, 2011; Kvesitadze et al., 2012). Compared with the other processes, dark fermentation is not only a way to generate energy but also to treat the waste. Many kinds of wastewater and solid waste from food industry and agriculture could be used as feedstock for biohydrogen production through dark fermentation (Wang and Wan, 2009). In addition, dark fermentation is much faster than photo-fermentations and without requirement for a direct input of solar energy (Argun and Kargi, 2011; Kim and Kim, 2013). The aim of this chapter is mainly to present an updated overview of biohydrogen production through dark fermentation. Firstly, the biochemical pathway, kenitic modes and key factors of dark fermentation are reviwed. Then, dark fermentation combined with other process is also discussed. Finally, challenges and prospects of biohydrogen production through dark fermentation are outlined.

2. THE PATHWAY OF BIOHYDROGEN PRODUCTION THROUGH DARK FERMENTATION

The biohydrogen production through dark fermentation is a complex series of biochemical reactions by diverse group of bacteria. Figure 1 showed the major catabolic pathways involved in glucose fermentation in mixed cultures (Cai et al., 2011; Lee et al., 2008; Li and Fang, 2007). Glucose is firstly converted into pyruvate, producing adenosine triphosphate (ATP) from adenosine diphosphate (ADP) and the reduced form of nicotinamide adenine dinucleotide (NADH) via the glycolytic pathway. Pyruvate is then further converted into acetylcoenzyme A by two pathways. One is common in strict anaerobes (*Clostridium* sp.) and generates reduced ferredoxin (Fdred). Another generates formic acid (facultative anaerobes, *Enterobacter* and *Klebsiella*). Acetylcoenzyme A is finally converted into acetic acid, propionic acid, butyric acid, lactic acid, butanol and ethanol, as well as hydrogen, depending on the microbial species present and the prevailing conditions (Venetsaneas et al., 2009). There are stoichiometric equations for hydrogen fermentation based on the kind of volatile fatty acids (VFAs) and alcohols produced in the process.

$$C_6H_{12}O_6 + 6H_2O \rightarrow 12H_2 + 6CO_2 \tag{1}$$

$$C_6H_{12}O_6 + 2H_2O \rightarrow 2CH_3COOH + 4H_2 + 2CO_2 \tag{2}$$

$$C_6H_{12}O_6 + 2H_2O \rightarrow CH_3CH_2CH_2COOH + 2H_2 + 2CO_2 \tag{3}$$

$$C_6H_{12}O_6 + 2H_2O \rightarrow CH_3CH_2OH + CH_3COOH + 2H_2 + 2CO_2 \tag{4}$$

$$C_6H_{12}O_6 \rightarrow 2CH_3CH(OH)COOH \tag{5}$$

$$C_6H_{12}O_6 + 2H_2 \rightarrow 2CH_3CH_2COOH + 2H_2O \tag{6}$$

$$4H_2 + 2CO_2 \rightarrow CH_3COOH + 2H_2O \tag{7}$$

$$4H_2 + CO_2 \rightarrow CH_4 + 2H_2O \tag{8}$$

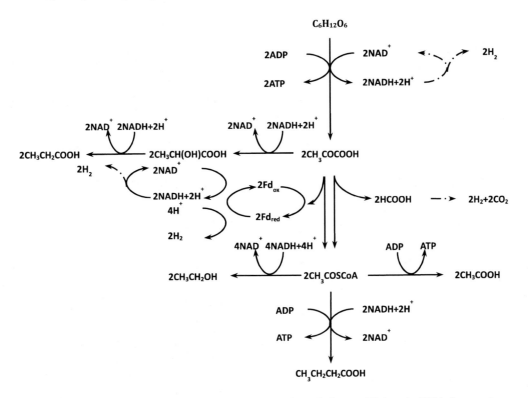

Figure 1. Pathway of hydrogen production via fermentation of glucose (Cai et al., 2011; Lee et al., 2008; Li and Fang, 2007).

Theoretically, 1mol glucose can produce 12 mol H_2 (Eq. 1). However, this reaction is energetically unfavorable with respect to biomass growth and would also only occur at extremely low hydrogen concentrations. There is a limit which called Thauer limit (Thauer et al., 1977), that at best, 1 mol glucose can only produce 4 mol H_2 (Eq. 2). The further fermentation of acetic acid cannot be achieved in dark conditions, but photo fermentation and microbial electrolytic cells.

Considering that the most common products in the fermentation of carbohydrate are acetic acid, butyric acid and ethanol (Li and Fang, 2007), the Eq. 2-4 were often used to described the biochemical reactions in the dark fermentation. Just as shown in the equations, the highest biohydrogen yield was coincident with butyrate accumulation, more ethanol correlated to reduced H_2 yield, and the H_2 yield was negligible when lactate was produced (Kim and Shin, 2008; Lee et al., 2008). The homoacetogenesis, methanogenesis and propionate production would consume the produced hydrogen (Eq. 6-8).

Based on the above biochemical reactions, a calculation of theoretical hydrogen (H_2 theoretical) in mmol production could be proposed (Arooj et al., 2008):

$$H_2 \text{ theoretical} = 2M_{But} + 2M_{Ac} - M_{Prop}$$

where M_{But}, M_{Ac} and M_{Prop} represented the butyric acid, acetic acid and propionic acid (mM), respectively.

The measured hydrogen production rate would be lower than stoichiometric values (Li and Fang, 2007). The highest yield achieved in strict anaerobes has been reported to be lower than 3 mol H_2/mol glucose in the literature (Cai et al., 2011) and the normal maximum H_2 yield in mesophilic biohydrogen fermentation is ~2 mol H_2/mol glucose (Lee et al., 2008). The possible reasons as follows: Firstly, glucose may be degraded through other pathways without producing hydrogen, such as direct acetic acid and lactic acid pathway.

Secondly, a fraction of glucose is consumed, instead, for biomass production. Thirdly, a stoichiometric yield is achievable only under near equilibrium condition, which implies a slow production rate and a low hydrogen partial pressure. The last but not the least, the produced hydrogen may be consumed though side reactions in the fermentation, such as methanogenesis and homoacetogenesis.

3. KINETIC MODEL FOR BIOHYDROGEN FERMENTION

In order to optimize the dark fermentation process, the mathematical models were used, including the Gompertz equation and anaerobic digestion model 1 (ADM1). The Gompertz equation is widely be used for kinetic modeling in the hydrogen batch experiment (De Gioannis et al., 2013; Ginkel et al., 2001; Lay et al., 1999; Peixoto et al., 2012). The equation as follows:

$$P = P_s \exp\left[-\exp\left(R_m \times e / P_s \times (\lambda - t) + 1\right)\right]$$

where P is the cumulative H_2 production (ml); P_s is H_2 production potential (ml); R_m is the maximum H_2 production rate (ml/h); e is the exp(1) = 2.71828; λ is the lag time (h); and t is the incubation time (h). For growth of hydrogen-producing bacteria, there is a modification equation for Gompertz equation, in the equation.

$$X = X_{max} \exp\left[-\exp\left(R_x \times e / X_{max} \times (\lambda - t) + 1\right)\right] + X_0$$

X is the bacteria concentration; X_{max} is the maximum net cell growth concentration (g/l); X_0 is the initial bacteria concentration (g/l); R_x is the maximum cell growth rate (g/lh); λ is the lag-phase time of bacteria growth (h); and t is the incubation time (h).

The ADM1 is also widely applicable model, which was designed for the modelling of anaerobic digestion of sewage sludge developed by international water association (IWA) task group in 2001. This model involves equations describing the biochemical processes (hydrolysis, acidogenesis, and methanogenesis) and chemical and physicochemical processes occurring during the fermentation (gas transfer, acid-base equilibrium). In order to describe the biohydrogen production through dark fermentation, the ADM1 was modified to contain lactate and ethanol among the metabolite. The validation of batch experiments showed that the modified ADM1 improved the prediction of fermentative hydrogen production (Antonopoulou et al., 2012). However, further modifications could be made in order to further improve the predictions for the hydrogenogenic process.

4. Factors Influencing Hydrogen Fermantation

As mentioned above, the biohydrogen production through dark femntations is a very complex biochemical process. It is influenced by many factors such as inoculum, pretreatment, pH, temperature, feedstock and reactors types. As shown in the Table 1, these crucial factors would be discussed in this section.

4.1. Inoculum

Some of the bacteria known to produce hydrogen included strict anaerobes (*Clostridiaceae*), facultative anaerobes (*Enterobactericeae* and *Klebsiella*) and even aerobes (*Bacillus, Aeromonons* spp., *Pseudomonos* spp. and *Vibrio* spp.)(Ginkel et al., 2001; Lee et al., 2011; Wang and Wan, 2009). Among these, *Clostridium* and *Enterobacter* were most widely used as microbes for fermentative hydrogen production (Wang and Wan, 2009).

Although the pure microbes for hydrogen production were broadly studied, the mix culture were recommended for the high robust and easy to get. In addition, the microorganisms other than main hydrogen producers (*Clostridium*, *Enterobacter*) would play important roles in in anaerobic fermentative biohydrogen production systems (Hung et al., 2011). For example, some facultative anaerobes such as *Streptococcus* sp. and *Klebsiella* sp. may be functioned as oxygen consumer to provide a strict anaerobic environment for hydrogen producer *Clostridium* (Hung et al., 2011). *Streptococcus* sp. could strengthen the architecture of the biological granule for hydrogen produce by forming a net-like structure with *Clostridium* (Hung et al., 2011). Some bacteria could increase hydrogen production from the breakdown of complex organic substrates such as cellulose (Elsharnouby et al., 2013).

The hydrogen-producing microbes could be achieved from different kinds of resources, as shown in the Table 1, including anaerobic sludge (AS), aerobic activated sludge (AAS), granular sludge (GS), soil, hydrogen-producing microorganism seed (HMS) and anaerobic sludge from hydrogen production reactor (ASH). In some cases, even though no inoculum the efficient hydrogen production was achieved (Wang and Zhao, 2009). In a word, hydrogen-producing bacteria can be easily obtained from the environment, and the most common inoculum is AS.

Table 1. Operating parameters and performance data for biohydrogen production from wastewater and solid waste

Feedstock	Inoculum Type	Pretreatment	Temperature °C	pH	Reactor Type	Operation	OLR (g COD/L/d)	HRT (h)	Specific H_2 Production	Reference
Glucose	NA	100°C, 15 min	38	5.7±0.2	UASB	Continuous 3.85	7	24	1.2[a]	(Alzate-Gaviria et al., 2007)
Sucrose	AS	100°C, 45 min	35	6.7±0.2	UASB	Continuous 3	20	4-24	0.75[a]	(Chang and Lin, 2004)
RWW	AS	acclimated	20-55	4.6-6.0	UASB	Continuous 3	14-36	2-24	1.37-2.14[a]	(Yu et al., 2002)
SIW	AAS	No treat	60	6.8	UASB	Continuous Lab	10.6-63.7	12-72	2.52[a]	(Ueno et al., 1996)
MSW	NA	100°C, 15 min	38	5.6	PBR	Batch 22.4			0.92[a]	(Alzate-Gaviria et al., 2007)
MSW	HMS	100°C, 15 min	37	5.2	-	Batch 0.12			180[b]	(Lay et al., 1999)
Sucrose	Soil	104°C, 2h	37	4.5-7.5	-	Batch 0.25	1.5-44.8	-	2.45[a]	(Ginkel et al., 2001)
CM	SWS	Ultrasonic 100 KHz,15 min	30-55	4.5-7.5	-	Batch 5	-	-	-	(Tang et al., 2008)
Glucose	AS	100°C,30 min & pH 10,24 h	-	6.5	UASB	Continuous 5	10-20	24,12	2.45[a]	(Liu et al., 2012)
Sucrose	AS	90°C,10 min & pH 3, 24 h	25	6.5	PBR	Continuous 2.5	24	2	1.05-0.35[a]	(Penteado et al., 2013)
Cheese whey	-	-	30	4.0-6.2	PBR	Continuous	22-37	24	0.55[a]	(Perna et al., 2013)
Glucose	AS	70°C, 30 min	37	6.5	PBR	Continuous	6.5-51.4	8	0.98-2.0[a]	(Júnior et al., 2014)
Sugarcane vinasse	HMS	-	55	6.5	PBR	Continuous	32.6-108.6	24-8	0.3-1.4[a]	(Júnior et al., 2014)
Sucrose	AS	No treat	25	6.5	PBR	Continuous 2.5	24	2	2.11[a]	(Lima and Zaiat, 2012)
Sucrose	AS	pH 3-4, 24 h	35	6.7	PBR	Continuous	240	0.5-5	0.10-1.32[d]	(Chang et al., 2002)
CM	AS	acclimated	60	5.2	CSTR	Semi-Continuous	31	10	10[b]	(Wang et al., 2013)
Sucrose	AS	pH 3, 24 h	35	6.7	CIGSB	Continuous 1	960-120	4-0.5	1.23[a]	(Lee et al., 2004)
Glucose	AS	acclimated	36	4-7	CSTR	Continuous 3	84	6	2.1[a]	(Fang and Liu, 2002)

Feedstock	Inoculum		Temperature °C	pH	Reactor		OLR (g COD/L/d)	HRT (h)	Specific H_2 Production	Reference
	Type	Pretreatment			Type	Operation				
HSW	AS	100 °C,1h	37	3.5-8.5	-	Batch 1				(Liu et al., 2006)
MSW	AS	acclimated	55	6	-	Batch 0.1			-	(Kobayashi et al., 2012)
CMS	AAS	100°C, 45 min	35	5.5	CSTR	Continuous 4	40-240	3-24		(Lay et al., 2010)
Glucose	AS		35	7		Batch				(Wang and Wan, 2008a)
FW	AS	100°C, 15 min	37	7.6-7.8	LBR	Batch		8		(Han et al., 2005)
Glucose	GS	acclimated	70	3.7	UASB	Continuous	25.1	5	0.73[a]	(Tähti et al., 2013)
FW	AS	no treat	55	5.5	CSTR	Continuous	64.4	31.2	205[b]	(Chu et al., 2008)
FW	No	-	40		CRD	Semi-Continuous	22.6 g VS/L/d	10	65[b]	(Wang and Zhao, 2009)
Cornstalks	AS	100°C, 15 min	37	6.5	-	Batch	50 g TS/L/d		63.7[c]	(Lu et al., 2009)
Cellulose	AS	no treat	37,55,80	5.7-5.9	CSTR	Continuous	5g TS/L	240	0.6-19.0[c]	(Gadow et al., 2012)
Glucose	AS	acclimated	37	5.5	AFBR	Continuous	5-120 g TS/L	0.125-3	0.4-1.7[a]	(Zhang et al., 2008)

MSW: municipal solid waste; RWW: rice winery wastewater; FW: food waste; SIW: sugar-industry waste; CM: cow manure; HSW: household solid waste; CMS: Soluble condensed molasses; SWS: sewage sludge; AS: anaerobic sludge; AAS: aerobic activated sludge; NA: non-anaerobic inoculum; HMS: Hydrogen-producing microorganism seed; a: mol H_2/mol hexose; b: mL/gVS; c: mL/g TS; d: L/L/d.

4.2. Pretreatment

The pretreatment of inoculum is achieved mostly by relying on the spore-forming characteristics of the hydrogen-producing *Clostridium* (Li and Fang, 2007). Pretreatment of inoculum can increase the performance of biohydrogen production and may also influence the VFAs, the ratio of acetic/butyric acids (de Sá et al., 2011).

The main pretreatment for inoculum methods reported in literature included heat treatment, acid shock, base shock, chemical pretreatment, ultrasonic treatment and combinations of the methods. Efforts have been made to compare the effect of pretreatment on the biohydrogen production. Penteado et al., investigated the influence of different pretreatment methods including heat treatment ($90^{\circ}C$, 10min) and acid shock (pH 3 for 24 hour then adjust to 7) on biohydrogen production in the up flow anaerobic fixed-bed reactors. The results showed that although heat treatment produced the maximum hydrogen value, acid treatment resulted in more stable hydrogen production with the largest average value (Penteado et al., 2013). Cheong et al., observed a similar result when investigated the pretreatments including acute physical and chemical ways. The results showed that acidification using perchloric acid had the best hydrogen production potential (Cheong and Hansen, 2006). However, some other studies drew the different conclusion that the heat treatment was more effective. Wang and Wan investigated the pretreatment of digested sludge by five methods (acid, base, heat-shock, aeration and chloroform) s to compare their suitability in the enrichment of hydrogen-producing bacteria. The result showed that the digested sludge pretreated by heat-shock could obtain the maximal hydrogen production potential, maximum hydrogen production rate, hydrogen yield, substrate degradation efficiency and biomass concentration (Wang and Wan, 2008a). The similar result was also reported by Argun et al., which compared three different pretreatment methods: repeated heat, chloroform and combinations of heat-chloroform treatment, and the highest hydrogen formation and specific hydrogen production rate were obtained with repeated heat pretreatment (Argun and Kargi, 2009). The study of Rossi et al., also reported that compared with acid, base and freezing and thawing, heat treatment is a simple and effective pretreatments method that can be used to improve the biotechnological production of hydrogen (Rossi et al., 2011).

These conflicts of studies may be resulted from the different inoculum they adopted. In Pendyala et al., study, both granular and flocculated anaerobic mixed cultures were pretreated using heat, shock loading, acid, alkali, linoleic acid and 2-bromoethan esulphonic acid (Pendyala et al., 2012). The hydrogen production performance showed that the kind of inoculum was the dominating factor but pretreatment method as flocculated controls contained a higher percentage of hydrogen producers in comparison to the granular cultures (Pendyala et al., 2012).

In general, the most common pretreatment is heat treatment and pH shock, appears to be simple and effective. However, there are a lot of studies reported that pretreatment could not inhibit the activity of all hydrogen-consuming bacteria (Lee et al., 2010; Li and Fang, 2007; Liu et al., 2006; Saady, 2013). This is because the characteristic of hydrogen-production microbes is not directly associated with the ability to form endospores. For instance, there are some non-spore-forming hydrogen producers include enteric bacteria like *Enterobacter* spp. and *Citrobacter* spp. There are also many hydrogen-consuming groups of bacteria that can form spores and could survive from heat treatment, such as acetogens (*Acetobacterium*, some

Clostridium spp., *Sporomusa*), certain propionate and lactate producers (*Propionibacterium, Sporolactobacillus*) (Chu et al., 2010).

4.3. pH

Control of pH was effective in changing the anaerobic pathway (Carrillo-Reyes et al.,, 2014; Kim et al., 2004). For example, butyrate accumulation was occurred with pH ~ 4, whereas lactate and propionate were dominant products at conditions close to neutral pH (Lee et al., 2008). Optimal pH can greatly improve the hydrogen fermentation performance. Low pH is one of the most critical factors to inhibit the activity of methanogenesis as the neutral pH is the optimal condition for methane production. As shown in Table 1, the most common pH ranges for biohydrogen production were: 5-5.5 and 6.5.

There were lots of studies about pH effect on biohydrogen production. In Chen et al. study, an anaerobic sequencing batch reactor was used to investigate the effects of pH (4.9, 5.5, 6.1, and 6.7) on biohydrogen production from carbohydrate-rich organic wastes. The results suggested that the maximum hydrogen yield appeared at pH 4.9 (Chen et al., 2009). Fang et al., investigated the effect of pH (4 to 7, 0.5 intervals) on the conversation of glucose to hydrogen by a mixed culture of fermentative bacteria, and found that the optimal pH is 5.5 (Fang and Liu, 2002). However, some batch experiment conducted in the neutral pH (Cheong and Hansen, 2006; Wang and Wan, 2008a).

The conflicts may be owed the different methods of pH control, that is, the initial control and process control. Lee drew the conclusion that hydrogen production was not controlled by initial pH when the final pH was low (Lee et al., 2008). Besides, the difference conclusion may be owed to the types of feedstock. For carbohydrates, its optimal pH range is acid (4.5-6), but for protein, it is alkaline pH (8.5-10) (Xiao et al., 2010). The optimal ranges of pH could be influenced by others conditions, such as organic loading rate (OLR) and mode of reactors running, that is, continuous or batch. Skonieczny et al., examined the effect of initial pH (range 5.7–6.5) and substrate loading (range 1-3 g COD/L) on the specific conversion and hydrogen production rate has shown interaction behavior between the two independent variables (Skonieczny and Yargeau, 2009). Highest conversion was achieved at pH of 6.1 and glucose concentration of 3 g COD/L, whereas the highest production rate was measured at pH 6.3 and substrate loading of 2.5 g COD/L (Skonieczny and Yargeau, 2009). Ginkel et al., studied the effect of varying initial pH (4.5 -7.5) and substrate concentration (1.5-44.8 g COD/L) on hydrogen production. The results indicated that the highest rate of hydrogen production occurred at a pH of 5.5 and a substrate concentration of 7.5 g COD/L (Ginkel et al., 2001).

4.4. Temperature

As shown in the Table 1, the most common temperature range of hydrogen fermentation were 20-25°C, 35-39°C, 40-60°C and > 60°C, corresponding to ambient fermentation, mesospheric fermentation, thermophilic fermentation and hyper-thermophilic fermentation, respectively.

Tang et al., invested the effect of temperatures on for biohydrogen production from cattle wastewater. The experiments were performed at various fermentation temperature values from 30°C to 50°C with testing at 0.5°C intervals, and the results showed that the optimal temperature was 45°C (Tang et al., 2008). The similar result was also reported by Wang et al., in which study showed that the optimized temperature for fermentative hydrogen production was 40°C (20°C to 55°C) (Wang and Wan, 2008b). The higher temperature range was also studied (37°C, 60°C, 70°C), and the results showed that the optimal temperature fermentative hydrogen production from cassava stillage was 60°C (Luo et al., 2010). Gadow et al., found the hyper-thermophilic (80°C) and thermophilic temperature (55°C) would inhabit the methane production and get a higher hydrogen yield (19.0 and 15.2 mmol/g cellulose, respectively) than mesophilic (37°C, 0.6 mmol/g cellulose) in CSTR (Gadow et al., 2012). Valdez-Vazquez et al., found that both the higher hydrogen contents and yields were achieved under thermophilic (55°C) than mesospheric (35°C) in semi-continuous experiments (Valdez-Vazquez et al., 2005). In conclusion, the results above suggested that superiority of thermophilic hydrogen fermentation. However, the conflict results were also reported by Shi et al., which found the higher hydrogen yield in mesophilic condition (35°C, 61.3 ± 2.0 ml/g TS) than thermophilic (50°C, 49.7 ± 2.8 ml/g TS) and hyper-thermophilic (65°C, 48.1 ± 2.5 ml/g TS) conditions (Shi et al., 2013).

As for practical engineering, ambient temperature is economic and simple to conduct. Lin invested the hydrogen fermentation in ambient temperature without control, and the results showed stable performance characteristics of a hydrogen-producing fermenter (Lin and Chang, 2004). A hydrogen yield of 0.35-1.05 mol/mol hexose was also reported by Penteado et al., using a continuous reactor operated at 25°C (Penteado et al., 2013). Moreover, Perera et al., reported the advantages of ambient temperature fermentation as the net energy gain would decline with increase of temperature, and the net energy yields were negative when the fermentation temperature exceeded 25°C (Perera et al., 2010).

4.5. Feedstock

Diverse feedstocks were investigated in the hydrogen fermentation except for common model compound (glucose, sucrose and cellulose). As shown in the Table 1, not only solid waste such as sugar-industry waste (SIW), household solid waste (HSW), municipal solid waste (MSW), food waste (FW), cow manure (CM), potato waste (PW) and cornstalk waste (CS) but also wastewater such as rice winery wastewater (RWW), cheese whey (CW) and rice wastewater (RW) were used for biohydrogen production through dark fermentation.

There were different hydrogen fermentation performances as different feedstock. It is clear that carbohydrates are the main source of hydrogen, hence wastes rich in carbohydrates can be used as substrates (Venetsaneas et al., 2009), for example, food waste, straw and cow manure. Kobayashi studied the various municipal solid wastes have different composition and found that wastes rich in carbohydrate in general had higher hydrogen yield than those rich in protein and fat (Kobayashi et al., 2012). Common food wastes such as dining hall or restaurant waste and wastes generated from food processing industries have shown good percentages of hydrogen in gas composition, production yield and rate (Yasin et al., 2013).

Carbon to nitrogen ratio is another very important characteristic of feedstock. Argun et al., found the C/N/P ratio maximizing the yield and formation rate of hydrogen was

100/0.5/0.1 (Argun et al., 2008), this is similar to study of Lima et al., which found that optimal carbon to nitrogen (C/N) ratio is 140 (Lima and Zaiat, 2012)**.** However, according to Lin et al., results, a C/N ratio of 47 provides the optimal biohydrogen production based on the microflora ability to convert sucrose into hydrogen or the microflora hydrogen production rate (Lin and Lay, 2004a). Sreethawong et al., obtained best hydrogen fermentation performance at the C/N ratio about 17 (Sreethawong et al., 2010). It showed that the optimal C/N ratio has no unified conclusion yet. The possible reason for this disagreement was the difference among these studies in the terms of inoculum, substrate and pH range studied. Phosphate also a very important essential element for biohydrogen production, Lin et al., recommended that use phosphate instead of carbonate as a buffering capacity supplement might be a useful strategy for optimal hydrogen production operations with anaerobic sewage cultures (Lin and Lay, 2004b) .

Nutrients of feedstock are also an important factor. Lin et al., conducted a batch experiment to exploit nutrient formulation for biohydrogen production by anaerobic microflora, included Mg, Na, Zn, Fe, K, I, Co, NH_4^+, Mn, Ni, Cu, Mo and Ca. The results indicated that Mg, Na, Zn and Fe affected the hydrogen production in a concentration dependent way and Mg being the most significant factor (Lin, 2005). The supplementation of calcium ion was found to increase biomass concentration and hydrogen production rate in a granules system (Lee et al., 2004).

4.6. Reactors Types

There are many different reactors types using for biohydrogen production, as shown in the Table 1 and Figure 2, such as continuous stirred tank reactor (CSTR), leaching-bed reactor (LBR), continuous rotating drum (CRD), anaerobic sequencing batch reactor (ASBR), up flow anaerobic sludge blanket (UASB), packed blanket reactor (PBR), carrier-induced granular sludge bed (CIGSB) and anaerobic fluidized bed reactor (AFBR).

In a conventional CSTR, biomass is well suspended in the mixed liquor, which has the same composition as the effluent. Since biomass has the same retention time as the HRT, washout of biomass may occur at shorter HRT. In addition, biomass concentration in the mixed liquor and the hydrogen production is limited (Wang and Wan, 2008a). Despite of these problems, CSTR is suitable for solid fermentation, for example food waste and crops straw. If the feedstock mainly is liquid, then the high-rate fermentation reactors were recommended, such as UASB, PBR and GIGSB. High-rate reactors provide an alternative to a conventional CSTR, because they are capable of maintaining higher biomass concentrations by forming granules, biofilm, or entrapped bioparticle (Wang and Wan, 2009). These reactors could operate at shorter HRT without biomass washout. The AFBRs based on biofilm and granules systems were used for hydrogen fermentation, and the HRT could be shortened to 0.125-3 h with a maximum hydrogen production rate of 7.6 L/L/h (Zhang et al., 2008).The short HRT 0.5-2 h HRT also was applied to PBR by Chang et al., and the maximal hydrogen production of 1.32 L/L/h rate was achieved (Chang et al., 2002). Ueno et al., investigated hydrogen production from sugar industry waste using the UASB could obtain a hydrogen production yield of 2.52 mol H_2/ mol glucose at the HRT of 12 h (Ueno et al., 1996).

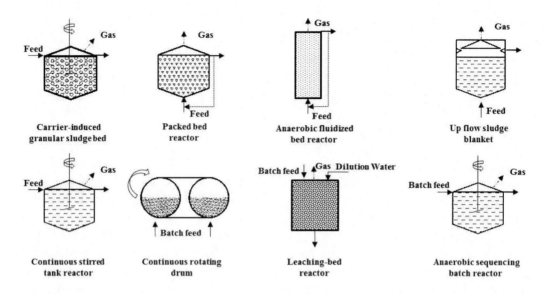

Figure 2. Reactor configurations for biohydrogen fermentation.

5. HYDROGEN FERMENTATION COMBINE WITH OTHER PROCESS

Biohydrogen production through dark fermentation does not significantly reduce the organic matter content of the substrates (Peixoto et al., 2012). The chemical oxygen demand (COD) removal is around 20% during the process (Luo et al., 2011). Accordingly, energy recovery is low during the process. The theoretical hydrogen production efficiency is only 33.5% as the Eq 2 (Xie et al., 2008). Hence, the effluents of hydrogen fermentation need to be further treated. There are a lot of studies of integrating hydrogen fermentation with other process, such as biomethane production (Demirel et al., 2010), microbial fuel cells (Oh and Logan, 2005; Sekoai and Gueguim Kana, 2014) and photo-fermentation (Huibo et al., 2009; Jun et al., 2012).

Table 2. Comparison of single-stage fermentation and two-stage fermentation

Feedstock	Single-stage fermentation	Two-stage fermentation		Increase in methane yield	Increase in energy yields	Reference
	Methane yield	Biohydrogen yield	Biomethane yield			
Wheat feed pellets	261 L/kg VS	7 L /kg VS	359 L/kg VS	37%	38%	(Massanet-Nicolau et al., 2013)
Cassava distillage	125 L/kgCOD$_{removed}$	-	147 L/kgCOD$_{removed}$	17.6%	-	(Zhang et al., 2013)
Nannochloropsis oceanica	171.3 L/kg VS	39 L/kg VS	114.2 L /kg VS	No	36%	(Xia et al., 2013)
Swine manure and market biowaste	404 L/kgVS$_{added}$	140 L/kgVS$_{added}$	351 L /kgVS$_{added}$	No	No	(Schievano et al., 2012)
Organic wastes	329 ± 13.7 L/kgVS	69 ± 6.3 L/kgVS	348 ± 14.2 L/kgVS	6%	11%	(Luo et al., 2011)

Among them, the combine of biohydrogen and biomethane production was regarded as a promising strategy (Liu et al., 2013). Compared with traditional single-stage fermentation for biomethane production, the two-stage could achieve a shorter HRT and higher OLR (Luo et al., 2011; Massanet-Nicolau et al., 2013).The methanogenic step in the two-stage process can be very stable with a large range of OLRs and a higher COD removal ratio and specific methane yield can be obtained (Zhang et al., 2013). As shown in the Table 2, the integration had a positive effect of both gas yields and energy yields. Furthermore, biohythane, a mixture of methane and hydrogen from the process, has been found to improve the performance of the combustion engine and can act as an important intermediate when switching from combustion engine to fuel-cell-powered electric vehicles (Liu et al., 2013).

CONCLUSION AND PERSPECTIVES

This chapter reviewed the pathways and models of the biohydrogen production through dark fermentation. Several main factors including inoculum, temperature, pH, feedstock, and reactor types were discussed and analyzed. Based on the advances in the researches, the optimal values of each factor for hydrogen fermentation were concluded. However, there were some conflicts in these studies. The possible reasons were discussed.

There are still several challenges to be solved although the promising potential of biohydrogen production through dark fermentation. One of the main challenges is low hydrogen yield. Several methods have been adopted to improve the hyrogen yield, including process modification, bioreactor design, genetic and molecular engineering technique. Another challenge is the high cost of the process, such as pretreatment of innoculum and feedstock which need heating or chemicals adding. Whst's more, the unstale hydrogen production process limited the popularity and applications for the dark fermentation. Tis is possibly attributed to the metabolic shift of hydrogen producing bacteria, and could be minimized by in depth study of the microbial community.

ACKNOWLEDGMENTS

This work was financially supported by Natural Science Foundation of China (21106080), the Chinese Universities Scientific Fund (2012RC030), the National Key Technology Support Program of China (2014BAD02B03), and NSFC-JST Cooperative Research Project (21161140328).

REFERENCES

[1] Alzate-Gaviria, L.M., Sebastian, P.J., Pérez-Hernández, A., Eapen, D., 2007. Comparison of two anaerobic systems for hydrogen production from the organic fraction of municipal solid waste and synthetic wastewater. *Int. J. Hydrogen Energ.*, 32, 3141–3146.

[2] Antonopoulou, G., Gavala, H.N., Skiadas, I.V., Lyberatos, G., 2012. Modeling of fermentative hydrogen production from sweet sorghum extract based on modified ADM1. *Int. J. Hydrogen Energ.*, 37, 191-208.

[3] Argun, H., Kargi, F., 2009. Effects of sludge pre-treatment method on bio-hydrogen production by dark fermentation of waste ground wheat. *Int. J. Hydrogen Energ.*, 34, 8543–8548.

[4] Argun, H., Kargi, F., 2011. Bio-hydrogen production by different operational modes of dark and photo-fermentation: An overview. *Int. J. Hydrogen Energ.*, 36, 7443–7459.

[5] Argun, H., Kargi, F., Kapdan, I.K., Oztekin, R., 2008. Biohydrogen production by dark fermentation of wheat powder solution: Effects of C/N and C/P ratio on hydrogen yield and formation rate . *Int. J. Hydrogen Energ.*, 33, 1813–1819.

[6] Arooj, M.F., Han, S., Kim, S., Kim, D., Shin, H., 2008. Continuous biohydrogen production in a CSTR using starch as a substrate . *Int. J. Hydrogen Energ.*, 33, 3289–3294.

[7] Balat, M., 2008. Potential importance of hydrogen as a future solution to environmental and transportation problems. *Int. J. Hydrogen Energ.*, 33, 4013-4029 .

[8] Cai, G., Jin, B., Monis, P., Saint, C., 2011. Metabolic flux network and analysis of fermentative hydrogen production. *Biotechnol. Adv.*, 29, 375–387.

[9] Carrillo-Reyes, J., Celis, L.B., Alatriste-Mondragón, F., Razo-Flores, E., 2014. Decreasing methane production in hydrogenogenic UASB reactors fed with cheese whey . *Biomass Bioenergy*, 63, 101–108.

[10] Chang, F., Lin, C., 2004. Biohydrogen production using an up-flow anaerobic sludge blanket reactor. *Int. J. Hydrogen Energ.*, 29, 33–39.

[11] Chang, J., Lee, K., Lin, P., 2002. Biohydrogen production with fixed-bed bioreactors. *Int. J. Hydrogen Energ.*, 27, 1167–1174.

[12] Chen, W., Sung, S., Chen, S., 2009. Biological hydrogen production in an anaerobic sequencing batch reactor: pH and cyclic duration effects. *Int. J. Hydrogen Energ.*, 34, 227–234.

[13] Cheong, D., Hansen, C., 2006. Bacterial stress enrichment enhances anaerobic hydrogen production in cattle manure sludge. *Appl. Microbiol. Biotechnol.*, 72, 635-643.

[14] Chu, C., Ebie, Y., Xu, K., Li, Y., Inamori, Y., 2010. Characterization of microbial community in the two-stage process for hydrogen and methane production from food waste. *Int. J. Hydrogen Energ.*, 35, 8253–8261.

[15] Chu, C., Li, Y., Xu, K., Ebie, Y., Inamori, Y., Kong, H., 2008. A pH- and temperature-phased two-stage process for hydrogen and methane production from food waste . *Int. J. Hydrogen Energ.*, 33, 4739–4746.

[16] De Gioannis, G., Muntoni, A., Polettini, A., Pomi, R., 2013. A review of dark fermentative hydrogen production from biodegradable municipal waste fractions. *Waste Manage.*, 33, 1345-1361.

[17] de Sá, L.R.V., de Oliveira, M.A.L., Cammarota, M.C., Matos, A., Ferreira-Leitão, V.S., 2011. Simultaneous analysis of carbohydrates and volatile fatty acids by HPLC for monitoring fermentative biohydrogen production. *Int. J. Hydrogen Energ.*, 36, 15177–15186.

[18] Demirel, B., Scherer, P., Yenigun, O., Onay, T.T., 2010. Production of methane and hydrogen from biomass through conventional and high-rate anaerobic digestion processes. *Crit. Rev. Env. Sci. Technol.*, 40, 116-146.

[19] Djomo, S.N., Blumberga, D., 2011. Comparative life cycle assessment of three biohydrogen pathways. *Bioresource Technol.*, 102, 2684-2694 .

[20] Dufour, J., Serrano, D.P., Galvez, J.L., Moreno, J., Garcia, C., 2009. Life cycle assessment of processes for hydrogen production. Environmental feasibility and reduction of greenhouse gases emissions. *Int. J. Hydrogen Energ.* 34 , 1370-1376 .

[21] Elsharnouby, O., Hafez, H., Nakhla, G., El Naggar, M.H., 2013. A critical literature review on biohydrogen production by pure cultures. *Int. J. Hydrogen Energ.*, 38, 4945 - 4966.

[22] Fang, H.H.P., Liu, H., 2002. Effect of pH on hydrogen production from glucose by a mixed culture . *Bioresource Technol.*, 82, 87–93.

[23] Gadow, S.I., Li, Y.Y., Liu, Y.Y., 2012. Effect of temperature on continuous hydrogen production of cellulose. *Int. J. Hydrogen Energ.*, 37 , 15465-15472 .

[24] Ginkel, S.V., Sung, S., Lay, J.J., 2001. Biohydrogen production as a function of pH and substrate concentration. *Environ. Sci. Technol.*, 35, 4726-4730.

[25] Han, S., Kim, S., Shin, H., 2005. UASB treatment of wastewater with VFA and alcohol generated during hydrogen fermentation of food waste. *Process Biochem.*, 40, 2897–2905.

[26] Huibo, S., Jun, C., Junhu, Z., Wenlu, S., Kefa, C., 2009. Combination of dark- and photo-fermentation to enhance hydrogen production and energy conversion efficiency. *Int. J. Hydrogen Energ.*, 34, 8846-53.

[27] Hung, C., Cheng, C., Guan, D., Wang, S., Hsu, S., Liang, C., Lin, C., 2011. Interactions between Clostridium sp. and other facultative anaerobes in a self-formed granular sludge hydrogen-producing bioreactor . *Int. J. Hydrogen Energ.*, 36, 8704–8711.

[28] Hung, C.H., Chang, Y.T., Chang, Y.J., 2011. Roles of microorganisms other than Clostridium and Enterobacter in anaerobic fermentative biohydrogen production systems-A review. *Bioresour Technol*, 102, 8437-44.

[29] Jun, C., Ao, X., Yaqiong, L., Richen, L., Junhu, Z., Kefa, C., 2012. Combination of dark- and photo-fermentation to improve hydrogen production from Arthrospira platensis wet biomass with ammonium removal by zeolite. *Int. J. Hydrogen Energ.*, 37, 13330-7.

[30] Júnior, A.D.N.F., Wenzel, J., Etchebehere, C., Zaiat, M., 2014. Effect of organic loading rate on hydrogen production from sugarcane vinasse in thermophilic acidogenic packed bed reactors . *Int. J. Hydrogen Energ.*, 39, 16852–16862.

[31] Júnior, A.D.N.F., Zaiat, M., Gupta, M., Elbeshbishy, E., Hafez, H., Nakhla, G., 2014. Impact of organic loading rate on biohydrogen production in an up-flow anaerobic packed bed reactor (UAnPBR). *Bioresource Technol.*, 164, 371–379.

[32] Kim, D.H., Kim, M.S., 2013. Development of a novel three-stage fermentation system converting food waste to hydrogen and methane. *Bioresource Technol.*, 267-274 .

[33] Kim, I.S., Hwang, M.H., Jang, N.J., Hyun, S.H., Lee, S.T., 2004. Effect of low pH on the activity of hydrogen utilizing methanogen in bio-hydrogen process . *Int. J. Hydrogen Energ.*, 29, 1133–1140.

[34] Kim, S., Shin, H., 2008. Effects of base-pretreatment on continuous enriched culture for hydrogen production from food waste . *Int. J. Hydrogen Energ.*, 33, 5266–5274.

[35] Kobayashi, T., Xu, K., Li, Y., Inamori, Y., 2012. Evaluation of hydrogen and methane production from municipal solid wastes with different compositions of fat, protein, cellulosic materials and the other carbohydrates. *Int. J. Hydrogen Energ.*, 37, 15711–15718.

[36] Kvesitadze, G., Sadunishvili, T., Dudauri, T., Zakariashvili, N., Partskhaladze, G., Ugrekhelidze, V., Tsiklauri, G., Metreveli, B., Jobava, M., 2012. Two-stage anaerobic process for bio-hydrogen and bio-methane combined production from biodegradable solid wastes. *Energy*, 37, 94–102.

[37] Lay, C.H., Wu, J.H., Hsiao, C.L., Chang, J.J., Chen, C.C., 2010. Biohydrogen production from soluble condensed molasses fermentation using anaerobic fermentation. *Int. J. Hydrogen Energ.*, 35 , 13445-13451 .

[38] Lay, J., Lee, Y., Noike, T., 1999. Feasibility of biological hydrogen production from organic fraction of municipal solid waste. *Water Res.*, 33, 2579–2586.

[39] Lee, D., Ebie, Y., Xu, K., Li, Y., Inamori, Y., 2010. Continuous H2 and CH4 production from high-solid food waste in the two-stage thermophilic fermentation process with the recirculation of digester sludge. *Bioresource Technol.*, 101, S42–S47.

[40] Lee, D., Show, K., Su, A., 2011. Dark fermentation on biohydrogen production: Pure culture. *Bioresource Technol.*, 102, 8393 - 8402.

[41] Lee, H., Salerno, M.B., Rittmann, B.E., 2008. Thermodynamic evaluation on H2 production in glucose fermentation. *Environ. Sci. Technol.*, 42, 2401–2407.

[42] Lee, K., Lo, Y., Lo, Y., Lin, P., Chang, J., 2004. Operation strategies for biohydrogen production with a high-rate anaerobic granular sludge bed bioreactor. *Enzyme Microb. Tech.*, 35, 605-612.

[43] Lee, K.S., Lo, Y.S., Lo, Y.C., Lin, P.J., Chang, J.S., 2004. Operation strategies for biohydrogen production with a high-rate anaerobic granular sludge bed bioreactor. *Enzyme Microb. Tech.*, 35, 605–612.

[44] Li, C., Fang, H.H.P., 2007. Fermentative hydrogen production from wastewater and solid wastes by mixed cultures. *Crit. Rev. Env. Sci. Technol.*, 37 , 1-39.

[45] Lima, D.M.F., Zaiat, M., 2012. The influence of the degree of back-mixing on hydrogen production in an anaerobic fixed-bed reactor. *Int. J. Hydrogen Energ.*, 37, 9630–9635.

[46] Lin, C., Chang, R., 2004. Fermentative hydrogen production at ambient temperature . *Int. J. Hydrogen Energ.*, 29, 715–720.

[47] Lin, C.Y., 2005. A nutrient formulation for fermentative hydrogen production using anaerobic sewage sludge microflora. *Int. J. Hydrogen Energ.*, 30, 285-292 .

[48] Lin, C.Y., Lay, C.H., 2004a. Carbon/nitrogen-ratio effect on fermentative hydrogen production by mixed microflora . *Int. J. Hydrogen Energ.*, 29, 41–45.

[49] Lin, C.Y., Lay, C.H., 2004b. Effects of carbonate and phosphate concentrations on hydrogen production using anaerobic sewage sludge microflora . *Int. J. Hydrogen Energ.*, 29, 275–281.

[50] Lin, C.Y., Lay, C.H., Sen, B., Chu, C.Y., Kumar, G., Chen, C.C., Chang, J.S., 2012. Fermentative hydrogen production from wastewaters: A review and prognosis. *Int. J. Hydrogen Energ.*, 37, 15632-15642 .

[51] Liu, D., Liu, D., Zeng, R.J., Angelidaki, I., 2006. Hydrogen and methane production from household solid waste in the two-stage fermentation process . *Water Res.*, 40, 2230–2236.

[52] Liu, Z., Lv, F., Zheng, H., Zhang, C., Wei, F., Xing, X., 2012. Enhanced hydrogen production in a UASB reactor by retaining microbial consortium onto carbon nanotubes (CNTs). *Int. J. Hydrogen Energ.*, 37, 10619–10626.

[53] Liu, Z., Zhang, C., Lu, Y., Wu, X., Wang, L., Wang, L., Han, B., Xing, X., 2013. States and challenges for high-value biohythane production from waste biomass by dark fermentation technology. *Bioresource Technol.*, 292-303.

[54] Lu, Y., Lai, Q., Zhang, C., Zhao, H., Ma, K., Zhao, X., Chen, H., Liu, D., Xing, X., 2009. Characteristics of hydrogen and methane production from cornstalks by an augmented two- or three-stage anaerobic fermentation process. *Bioresource Technol.*, 100, 2889–2895.

[55] Luo, G., Xie, L., Zhou, Q., Angelidaki, I., 2011. Enhancement of bioenergy production from organic wastes by two-stage anaerobic hydrogen and methane production process. *Bioresource Technol.*, 102, 8700–8706.

[56] Luo, G., Xie, L., Zou, Z., Zhou, Q., Wang, J., 2010. Fermentative hydrogen production from cassava stillage by mixed anaerobic microflora: Effects of temperature and pH. *Appl. Energ.*, 87, 3710–3717.

[57] Massanet-Nicolau, J., Dinsdale, R., Guwy, A., Shipley, G., 2013. Use of real time gas production data for more accurate comparison of continuous single-stage and two-stage fermentation. *Bioresource Technol.*, 129, 561-567 .

[58] Oh, S.E., Logan, B.E., 2005. Hydrogen and electricity production from a food processing wastewater using fermentation and microbial fuel cell technologies. *Water Res.*, 39, 4673-4682.

[59] Peixoto, G., Pantoja-Filho, J., Agnelli, J., Barboza, M., Zaiat, M., 2012. Hydrogen and Methane Production, Energy Recovery, and Organic Matter Removal from Effluents in a Two-Stage Fermentative Process. *Appl. Biochem. Biotechnol.*, 168, 651-671.

[60] Pendyala, B., Chaganti, S.R., Lalman, J.A., Shanmugam, S.R., Heath, D.D., Lau, P.C.K., 2012. Pretreating mixed anaerobic communities from different sources: Correlating the hydrogen yield with hydrogenase activity and microbial diversity. *Int. J. Hydrogen Energ.*, 37, 12175.

[61] Penteado, E.D., Lazaro, C.Z., Sakamoto, I.K., Zaiat, M., 2013. Influence of seed sludge and pretreatment method on hydrogen production in packed-bed anaerobic reactors. *Int. J. Hydrogen Energ.*, 38, 6137–6145.

[62] Perera, K.R.J., Ketheesan, B., Gadhamshetty, V., Nirmalakhandan, N., 2010. Fermentative biohydrogen production: Evaluation of net energy gain . *Int. J. Hydrogen Energ.*, 35, 12224–12233.

[63] Perna, V., Castello, E., Wenzel, J., Zampol, C., Lima, D., Borzacconi, L., Varesche, M.B., Zaiat, M., Etchebehere, C., 2013. Hydrogen production in an upflow anaerobic packed bed reactor used to treat cheese whey. *Int. J. Hydrogen Energ.*, 38 , 54-62 .

[64] Rossi, D.M., Costa, J.B.D., de Souza, E.A., Peralba, M.D.C.R., Samios, D., Ayub, M.A.Z., 2011. Comparison of different pretreatment methods for hydrogen production using environmental microbial consortia on residual glycerol from biodiesel. *Int. J. Hydrogen Energ.*, 36, 4814–4819.

[65] Saady, N.M.C., 2013. Homoacetogenesis during hydrogen production by mixed cultures dark fermentation: Unresolved challenge . *Int. J. Hydrogen Energ.*, 38, 13172–13191.

[66] Schievano, A., Tenca, A., Scaglia, B., Merlino, G., Rizzi, A., Daffonchio, D., Oberti, R., Adani, F., 2012. Two-Stage vs Single-Stage Thermophilic Anaerobic Digestion: Comparison of Energy Production and Biodegradation Efficiencies. *Environ. Sci. Technol.*, 46, 8502-8510 .

[67] Sekoai, P.T., Gueguim Kana, E.B., 2014. Semi-pilot scale production of hydrogen from Organic Fraction of Solid Municipal Waste and electricity generation from process effluents. *Biomass Bioenergy*, 60, 156-163.

[68] Shi, X., Kim, D., Shin, H., Jung, K., 2013. Effect of temperature on continuous fermentative hydrogen production from Laminaria japonica by anaerobic mixed cultures. *Bioresource Technol.*, 225–231.

[69] Show, K.Y., Lee, D.J., Tay, J.H., Lin, C.Y., Chang, J.S., 2012. Biohydrogen production: Current perspectives and the way forward. *Int. J. Hydrogen Energ.*, 37, 15616–15631.

[70] Skonieczny, M.T., Yargeau, V., 2009. Biohydrogen production from wastewater by Clostridium beijerinckii: Effect of pH and substrate concentration. *Int. J. Hydrogen Energ.*, 34, 3288–3294.

[71] Sreethawong, T., Chatsiriwatana, S., Rangsunvigit, P., Chavadej, S., 2010. Hydrogen production from cassava wastewater using an anaerobic sequencing batch reactor: Effects of operational parameters, COD:N ratio, and organic acid composition . *Int. J. Hydrogen Energ.*, 35, 4092–4102.

[72] Tähti, H., Kaparaju, P., Rintala, J., 2013. Hydrogen and methane production in extreme thermophilic conditions in two-stage (upflow anaerobic sludge bed) UASB reactor system. *Int. J. Hydrogen Energ.*, 38, 4997–5002.

[73] Tang, G.L., Huang, J., Sun, Z.J., Tang, Q.Q., Yan, C.H., Liu, G.Q., 2008. Biohydrogen production from cattle wastewater by enriched anaerobic mixed consortia: Influence of fermentation temperature and pH. *Biotechnol. Bioeng.*, 106, 80-87.

[74] Thauer, R.K., Jungermann, K.A., Decker, K., 1977. Energy conservation in chemotrophic anaerobic bacteria. *Bacteriological Rev.*, 41`.

[75] Ueno, Y., Otsuka, S., Morimoto, M., 1996. Hydrogen production from industrial wastewater by anaerobic microflora in chemostat culture. *J. Ferment. Bioeng.*, 82, 194–197.

[76] Valdez-Vazquez, I., Ríos-Leal, E., Esparza-García, F., Cecchi, F., Poggi-Varaldo, H.M., 2005. Semi-continuous solid substrate anaerobic reactors for H_2 production from organic waste: Mesophilic versus thermophilic regime. *Int. J. Hydrogen Energ.*, 30, 1383–1391.

[77] Venetsaneas, N., Antonopoulou, G., Stamatelatou, K., Kornaros, M., Lyberatos, G., 2009. Using cheese whey for hydrogen and methane generation in a two-stage continuous process with alternative pH controlling approaches. *Bioresource Technol.*, 100, 3713–3717.

[78] Wang, J., Wan, W., 2008a. Comparison of different pretreatment methods for enriching hydrogen-producing bacteria from digested sludge. *Int. J. Hydrogen Energ.*, 33, 2934–2941.

[79] Wang, J., Wan, W., 2008b. Effect of temperature on fermentative hydrogen production by mixed cultures . *Int. J. Hydrogen Energ.*, 33, 5392–5397.

[80] Wang, J., Wan, W., 2009. Factors influencing fermentative hydrogen production: A review . *Int. J. Hydrogen Energ.*, 34, 799–811.

[81] Wang, K., Chen, J., Huang, Y., Huang, S., 2013. Integrated Taguchi method and response surface methodology to confirm hydrogen production by anaerobic fermentation of cow manure. *Int. J. Hydrogen Energ.*, 38, 45–53.

[82] Wang, X., Zhao, Y., 2009. A bench scale study of fermentative hydrogen and methane production from food waste in integrated two-stage process . *Int. J. Hydrogen Energ.*, 34, 245–254.

[83] Xia, A., Cheng, J., Lin, R.C., Lu, H.X., Zhou, J.H., Cen, K.F., 2013. Comparison in dark hydrogen fermentation followed by photo hydrogen fermentation and methanogenesis between protein and carbohydrate compositions in Nannochloropsis oceanica biomass. *Bioresource Technol.*, 204-213 .

[84] Xiao, B., Han, Y., Liu, J., 2010. Evaluation of biohydrogen production from glucose and protein at neutral initial pH. *Int. J. Hydrogen Energ.*, 35, 6152–6160.

[85] Xie, B., Cheng, J., Zhou, J., Song, W., Liu, J., Cen, K., 2008. Production of hydrogen and methane from potatoes by two-phase anaerobic fermentation . *Bioresource Technol.*, 99, 5942–5946.

[86] Yasin, N.H.M., Mumtaz, T., Hassan, M.A., Rahman, N.A., 2013. Food waste and food processing waste for biohydrogen production: A review. *J. Environ. Manage.*, 130, 375–385.

[87] Yu, H., Zhu, Z., Hu, W., Zhang, H., 2002. Hydrogen production from rice winery wastewater in an upflow anaerobic reactor by using mixed anaerobic cultures. *Int. J. Hydrogen Energ.*, 27, 1359–1365.

[88] Zhang, Q., Zhu, X., Kong, L., Yuan, G., Zhai, Z., Liu, H., Guo, X., 2013. Comparative assessment of the methanogenic steps of single and two-stage processes without or with a previous hydrolysis of cassava distillage. *Bioresource Technol.*, 147, 1–6.

[89] Zhang, Z., Show, K., Tay, J., Liang, D.T., Lee, D., 2008. Biohydrogen production with anaerobic fluidized bed reactors—A comparison of biofilm-based and granule-based systems. *Int. J. Hydrogen Energ.*, 33, 1559–1564.

In: Gas Biofuels from Waste Biomass
Editor: Zhidan Liu

ISBN: 978-1-63483-192-5
© 2015 Nova Science Publishers, Inc.

Chapter 7

BIOHYDROGEN PRODUCTION VIA MICROBIAL ELECTROLYSIS CELL: A REVIEW

Ruixia Shen and Zhidan Liu[*]

Laboratory of Environment-Enhancing Energy (E2E),
College of Water Resources and Civil Engineering,
China Agricultural University, Beijing, China

ABSTRACT

Hydrogen production through microbial electrolysis cell (MEC) is one of the promising approaches. In this chapter, advances on biohydrogen production via MEC were reviewed, and the key factors affecting hydrogen production were discussed, including reactor configuration, electrode material, cathode catalyst, substrate, and auxiliary voltage. Specifically, operation parameters to assess the performance of MEC were commented, including chemical oxygen demand (COD) removal, current density (I_A), hydrogen production rate (Q_{H2}), coulombic efficiency (C_E), hydrogen recovery (R_{H2}) and energy recovery (η_E). Challenges and prospectives of biohydrogen production through MEC were finally addressed.

Keywords: biohydrogen, microbial electrolysis cell, key factors, operation parameters

1. INTRODUCTION

Since the 1970s, hydrogen has been produced as a kind of secondary energy through anaerobic fermentation (Yokoi et al., 2002; Tanisho et al., 1998; Lu et al., 2004). Traditional hydrogen generation technologies mainly include hydrogen generation through fermentation, hydrogen generation through electrolysis of water, pyrolysis of methane and so on, which all need to consume large amounts of non-renewable energy. This cannot adapt to the demand of sustainable development strategy and the trend of future energy. Biological hydrogen

---
[*] Email: zhidanliu@cau.edu.cn

production has attracted widespread attention in the world because of its obvious advantages: high hydrogen production rate, simple reactor design, hydrogen obtained in the short term, renewable biomass (organic waste and animal manure) and even high concentration organic wastewater from some industries (the agroforestry waste straw, firewood, food processing, brewing, livestock and poultry production), used as raw materials (Xing et al., 2005), and no serious secondary pollutant from microorganisms. In general, bio-hydrogen production can greatly reduce the demand on fossil energy from the perspective of energy conversion. MEC, as a new biological technology, originated from microbial fuel cell (MFC) (Logan et al., 2008), and consists of anode chamber and cathode chamber. It is a typical subject involving a variety of disciplines including biology, chemistry, materials, physics, process control, and telecommunication engineering. The anode environment and the anodic reaction in both MEC and MFC are consistent, so the anode material in MFC can be also used in MEC (Logan et al., 2006). Microorganisms are treated as the catalyst of anode substrate, the anode substrate thoroughly oxidized and decomposed into protons and electrons, then protons and electrons combined to produce clean energy (hydrogen) under lower auxiliary voltage (Douglas et al., 2008). The MEC is developed on the basis of MFCs, which can realize the double benefits of waste water treatment and hydrogen generation. In theory, a maximum of 12 mol hydrogen can be obtained from 1 mol glucose, but because of the existence of the thermodynamic limit of hydrogen generation from fermentation (Cheng et al., 2007), most of the substrate is converted into small organic molecules not hydrogen, which reduces the production efficiency of hydrogen. In addition, consumption of hydrogen from methanogens also affects the production efficiency of hydrogen (Hawkes et al., 2002; Kim et al., 2004; Oh et al., 2003), therefore in the actual hydrogen generation from fermentation, 1 mol glucose can only produce 2-3 mol hydrogen (Nath et al., 2004). In addition, the voltage of hydrogen generation through electrolysis of water is 1.23 V, while the voltage of hydrogen production by MEC is 0.114 V (Liu et al., 2005; Rozendal et al., 2006), and MEC can produce hydrogen with the help of a low auxiliary voltage. Hydrogen generation from chemical methods need to use fossil fuels, so it is imperative to replace the way of hydrogen generation from chemical methods with the continuous depletion of fossil fuels. In conclusion, hydrogen generation via the MEC has the advantages of high efficiency, energy saving, and environmental protection. In 2005, Liu et al., first reported double chamber MEC to produce hydrogen with sodium acetate as the substrate (Liu et al., 2005); In 2011, Roland explored the MEC from brewery wastewater in half industrial scale, providing the possibility of MECs toward industrialization (Roland et al., 2011).

Then, in recent years further studies were reported as showed in table 1. In this table, it includes some key factors affecting the production hydrogen of MEC, like the reactors types (single chamber; two-chamber), operation mode (fed-batch; continuous), substrate (kinds of wastewater; organic small molecule acid), electrode material (based-carbon material), catalysts (Pt), and auxiliary voltage (0.25-1.10 V). These key factors were also the research focus of this field for scholars in China and abroad. Additionally, to evaluate the performance of MEC operation, a few of indexes were determined including chemical indexes (COD removal; Q_{H2}; C_E; R_{H2}; η_E) and electrochemical indexes (I_A).

Table 1. The research status on MECs in recent years

Characteristics of reactors							Parameters		Evaluating indexes						References
Types of reactor	Anode	Cathode/ Catalyst	Substrate	Membrane	V (mL)	Operation types	E_{ap}	R_{ap}	COD removal (%)	I_A (A/m^2)	Q_{H2} ($m^3/m^3/d$)	C_E (%)	R_{H2} (%)	η_E (%)	
Two-chamber	Graphite granules	Graphite granules	Wheat straw	CEM	500	Fed-batch	0.7	—	61	—	0.61	—	64	78	(Anders et al., 2011)
Single chamber	Graphite plates	Stainless steel mesh	Refinery wastewater	—	5	Fed-batch	0.7	10	79	2.1	1.1	—	—	288	(Ren et al., 2013)
Two-chamber	Graphite granules	Carbon cloth/Pt	Acetic acid	AEM	14	Fed-batch	0.6	1000	—	0.72-0.75	—	—	91	261	(Cheng et al., 2007)
Two-chamber	Carbon paper	Carbon paper/Pt	Domestic wastewater	CEM	292	Fed-batch	0.5	1400	87-100	0.471	0.0125[a]	23.2	—	—	(Jenna et al., 2007)
Single chamber	Graphite-fiber	Carbon cloth/Pt	Bovine serum albumin	—	26	Fed-batch	0.6	10	86-91	—	0.42	—	—	75	(Lu et al., 2010)
Single chamber	Carbon cloth	Carbon cloth/Pt	Sodium acetate	—	5	Fed-batch	0.7	—	—	160 A/m^3	1.9	—	—	—	(Douglas et al., 2009)
Single chamber	Graphite brush	Carbon cloth/Pt	Waste activated sludge	—	28	Fed-batch	0.8	10	>60	—	0.1071[a]	—	—	155	(Liu et al., 2012)
Two-chamber	Plain carbon cloth	Carbon paper/Pt	Acetate	PEM	200	Fed-batch	0.25 - 0.85	—	>95	0.15-0.88	—	60-78	—	—	(Liu et al., 2005)
Single chamber	Graphite felt	Ni-based gas diffusion electrode	Domestic wastewater	—	90	Continuous	1	16	67	—	—	65	—	—	(Escapa et al., 2012)
Two-chamber	Carbon paper	Carbon cloth/Pt	Cellulose	AEM	28	Fed-batch	0.3	1000	68	—	—	27	—	—	(Cheng et al., 2011)
Two-chamber	Carbon felt	Titanium plate/Pt	Acetate	PEM	200	Fed-batch	0.5	600	—	—	—	64.6	61.1	—	(Chae et al., 2010)
Single chamber	Graphite fiber brush	Carbon cloth/Pt	Sodium acetate	—	28	Fed-batch	0.9	10	—	—	1.4	87	—	155	(Jack et al., 2011)
Single chamber	Carbon cloth	Carbon cloth /Pt	Sodium acetate	—	500	Semi-continuous	0.6	1.1	—	9.3	0.53	75	62	204	(Hu et al., 2008)
Two-chamber	Plain carbon cloth	Carbon paper/Pt	Acetate	—		Fed-batch	0.6	—	86	—	—	35	31	—	(Liu et al., 2012)

Table 1. (Continued)

Characteristics of reactors							Parameters		Evaluating indexes						References
Types of reactor	Anode	Cathode/Catalyst	Substrate	Membrane	V (mL)	Operation types	E_{ap}	R_{ap}	COD removal (%)	I_A (A/m²)	Q_{H2} (m³/m³/d)	C_E (%)	R_{H2} (%)	η_E (%)	
Two-chamber	Carbon brush	Stainless steel mesh/Pt	Domestic wastewater	AEM	28	Fed-batch	0.9	10	—	—	1.4	—	—	174	(Edgar et al., 2013)
Single chamber	Graphite brush	Carbon cloth/Pt	Glucose	—	26	Fed-batch	0.8	10	—	50 A/m³	0.37	74	51	128	(Lu et al., 2012)
Single chamber	Graphite fiber brush	Stainless steel 304 mesh #60	Carbonate	—	28	Fed-batch	1.05	—	—	—	2.3	>100	—	161	(Roland et al., 2012)
Single chamber	Graphite fiber brush	Stainless steel 304 mesh #60	Winery wastewater	—	1000	Continuous	0.9	0.01	62 (SCOD)	7.4 A/m³	0.06	—	—	—	(Roland et al., 2011)
Single chamber	Graphite plate	Stainless steel mesh	Domestic wastewater	—	5	Fed-batch	0.7	10	—	240 A/m³	—	—	—	—	(Call et al., 2011)
Single chamber	graphite brush	Carbon cloth/Pt	Acetate	—	28	Fed-batch	0.6	10	—	133 A/m³	1.1	—	—	187	(Nam et al., 2011)
Two-chamber	Plain carbon paper	Carbon paper/Pt	Acetate	PEM	450	Fed-batch	0.8	10	—	—	0.015	28-33	88-89	—	(Sun et al., 2008)
Single chamber	Graphite fiber brush	Carbon cloth/Pt	Sodium acetate	—	26	Fed-batch	1	—	—	1830 A/m³	17.8	—	93	115	(Cheng et al., 2011)
Single chamber	Graphite fiber brush	Graphite fiber cloth/Pt	Glycerol	—	28	Fed-batch	0.8	12	100	50 A/m³	—	35	—	—	(Nuria et al., 2015)
Two-chamber	Stainless steel mesh	Stainless steel wool	Domestic wastewater	RM	88	Continuous	1.1	0.1	29.8	—	0.007	41.2	—	48.7	(Elizabeth et al., 2014)
Two-chamber	Graphite brushes	Graphite brushes/SRB	Domestic wastewater	CEM	30	Fed-batch	0.8	10	—	—	—	50	—	—	(Luo et al., 2014)
Two-chamber	Carbon cloth	Air-diffusion/A aerobic sludge	Wastewater	CEM	137	Fed-batch	—	—	—	2.29	—	—	—	—	(Wang et al., 2013)

"–" in the table stands for no introduction in corresponding reference; "a" stands for "mg H$_2$/mg COD"; AEM stands for "anion exchange membrane"; CEM stands for "cation exchange membrane "; PEM stands for "proton exchange membrane "; RM stands for "Rhinohide membrane"; "E$_{ap}$" stands for applied voltage; "R$_{ap}$" stands for applied resistance; "V" stands for volume of MEC reactor; "SRB" stands for sulfate-reducing bacteria.

2. MAIN FACTORS AFFECTING HYDROGEN GENERATION VIA MECS

2.1. MEC Construction and Operation

Figure 1 showed some structure of MECs for hydrogen generation (Logan et al., 2008). MEC configurations usually were designed into two chambers (Jeremiasse et al., 2010; Rozendal et al., 2008; Chae et al., 2008). Since bacteria of the anode chamber are under anaerobic environment, nitrogen was usually used to drive out the residual oxygen around the anode. The anode and the cathode are separated by a membrane, like Proton-Exchange Membrane (Liu et al., 2005) and Cation-Exchange Membrane (Rozendal et al., 2008). Double chambers structure has many advantages: (1) Anodic reaction and cathodic hydrogen production reaction can be separated, which could inhibit the mutual diffusion between anode material and cathode material; (2) obtain relatively pure hydrogen; (3) greatly reduce the diffusion of hydrogen from the cathode to the anode, which could be consumed by the microorganisms in anode (exoelectrogens, homoacetogenic bacteria); (4) prevent the inactivation of the cathode catalyst from pollution. However, researchers have found that the membrane of two-chamber MEC cannot completely prevent hydrogen diffusion into the anode chamber. Then the membrane will affect the transfer of protons and hydroxyl ions between the two electrodes, which would lead to the pH gradient between two chambers (pH is reduced in the anode and increased in the cathode), the damage of exoelectrogens activity, and the higher internal resistance. Additionally, the increase of MEC internal resistance will directly result in a decline of current density and the hydrogen production rate, and an increase in energy consumption. Furthermore, it is difficult to amplify the two-chamber MEC because of its complexity of structure.

Figure 1. Some architectures of MECs. Four types used in fed-batch experiments: (A) H-type; (B and D) two cube-type; (C) cube-type single chamber; (E) Disc-shaped two-chamber; (F) disk- shaped; (G) rectangular-shaped.

Therefore, some researches have been done about the single chamber MEC because of its simple design, low cost and lower internal resistance (Call et al., 2008). Although hydrogen efficiency of single chamber MEC was improved, there are some drawbacks for the single chamber MEC, that is the purity of hydrogen is greatly restricted (a mixture of hydrogen and methane), and the content of hydrogen decreased because some microbes in anode (e.g., methanogen) can absorb some of hydrogen produced in cathode.

2.2. Catalyst

In MEC, the anode reduction is both the key step and the rate-limiting step for the whole energy conversion, so the catalyst (microbes) in the anode reaction became the research focus. In theory, most microorganisms are likely to be a biological catalyst in MEC. In 1987, Derek R. Lovley isolated *Geobacter metallireducens* from the river sediments of Potomac River (Derek et al., 1987). Subsequently, scientists have isolated many microbes, like *Shewanella putrefaciens*, *Geobacteraceae sulferreducens*, *Geobacter metallireducens*, and *Rhodoferax ferrireducens* (Kim et al., 2002; Bond et al., 2003; Min et al., 2005). These microbes possess the electrochemical activity, and can transfer electron from the substrate to the anode without any medium, finally form the biofilm. Even these microbes could stably operate in the reactor and achieve a high C_E. In recent years, there were a lot of researches about the biological catalyst on MEC, such as the separation of the electrochemical activity bacteria, the electrons generation, transmission, and transformation approaches in microbe, etc. Therefore, obtaining pure bacterium of high catalytic activity can not only improve the C_E of MEC, but also promote the development on electronic generation mechanism and transport pathway. However, currently mixed bacteria rather than pure ones are used as catalyst of anode, such as sewage sludge in wastewater treatment plant. The reason is that mixed bacteria have strong resistance to environmental impact, and widely available mechanism, so it is beneficial to the project amplification (Huang et al., 2007). In addition, the effect of substrate on microbial activity is also a focuses on microbiology, because different electron donor lead to the differences of the energy density (Liu et al., 2004; Liu et al., 2005).

Hydrogen evolution reaction happened in cathode, but the response is slow in the plain carbon electrode, needing over-potential drive reaction (Guo et al., 2010). Therefore, researchers in recent years mainly focused on the cathode catalyst, moreover cathode carrier material generally choose carbon based material, titanium wire, etc., making the catalyst attach to the carrier (Liu et al., 2005; Call et al., 2008; Rozendal et al., 2006). Transition metals are generally used as the catalyst due to their stable properties, such as Pt, Pd. In MEC, the cost of catalyst in cathode accounted for 47% of the total cost (Rozendal et al., 2008). In addition, Pd/Pt composite catalyst can effectively reduce the cost because Pd has a lower price, and studies have shown that Pd/Pt and Pt have almost the same hydrogen generation efficiency (Tartakovsky et al., 2008). Stainless steel and nickel alloy can also be used as effective cathode catalyst. Hydrogen generation performance using A286 stainless steel cathode is better than Pt cathode (Selembo et al., 2009). Compared to chemical catalysts, biological catalyst has more advantages, and it can reduce the cost and produce hydrogen continuously. In 2011, Elsemiek used the microbes from biological anode as biological cathode, and finally it obtained a hydrogen yield of 0.63 m^3/ ($m^3 \cdot d$) (Elsemiek et al., 2011).

Rozendal etc. found that microorganisms which has attached to the cathode may catalyze hydrogen evolution reaction. The results showed that hydrogen efficiency obtained in a double chamber MEC is 8 times than that of a common electrode, and the current density is 2.4 times than that of the platinum cathode under an auxiliary voltage of 0.7 V. Therefore, biological cathode can greatly reduce the hydrogen evolution potential compared to the Pt catalyst (Rozendal et al., 2008). Therefore, it has great significance to develop a biological MEC which have both biological anode and biological cathode (Liu et al., 2012).

2.3. Auxiliary Voltage

The auxiliary voltage is one of the main factors for hydrogen production efficiency of MECs, Furthermore, MECs require relatively low energy input (0.2-0.8 V) compared to typical water electrolysis (Liu et al., 2011; Zhang et al., 2014). Increasing the auxiliary voltage can shorten the reaction time and obtain higher hydrogen efficiency. However, too high voltage is not consistent with the principle of energy conservation. Conversely, if the voltage is too low, there is no hydrogen production. The appropriate voltage is determined by both the voltage of microbial growth in the reactor and the voltage which generated by the resistance of the system. Therefore, choosing the appropriate voltage is a key technology for hydrogen production.

In 2006, Rozendal R.A. etc. obtained 2.10 mol H_2/mol acetate using a larger MEC reactor (3.3 L) of two-chamber under the auxiliary voltage of 0.5 V, its $_{CE}$ is 92%, and the effect of the reactor can reach 0.02 $m^3 H_2/m^3$/d. In addition, the study has pointed out the methanogens existing in anode had effects on the hydrogen production, but it did not make an analysis from the perspective of the biological community (Rozendal et al., 2006). During the practical production process, MEC generate hydrogen in the initial stage, while methane was produced after long running of the mixed bacteria reactor. The competition phenomenon between methane production and hydrogen production in MEC is inevitable (Cusick et al., 2011). Lee's study showed that the production of methane is very obvious under a long-term operation of MEC without a membrane. The reactor operated in the initial sodium acetate concentration of 80mM for 7 days, and the methane concentration was 16% in the total gases (Lee et al., 2009).

In MEC system, the growth conditions of electron transfer function bacteria is extremely close to methanogens and the growth of methanogens result in the decrease of hydrogen efficiency. Therefore how to reduce or inhibit the growth of methanogens becomes the problem which we are faced with now in this field. Recent researches showed that the growth of methanogens could be controlled by changing the operating conditions, such as auxiliary voltage, temperature, pH, etc (Tartakovsky et al., 2008; Lu et al., 2011). But at the same time this may has some negative influences on the growth of electron transfer function bacteria. It is extremely difficult to guarantee both a higher hydrogen generation capacity and the controlling methane. Therefore, it was proposed that it was no need to focus on the control of methane, while it is promising to develop a kind of device about MEC which can produce both hydrogen and methane (Clauwaert et al., 2008).

In 2007, Logan's group used a variety of substrates, including glucose, cellulose, acetic acid, butyric acid, lactic acid, propionic acid, and valerian oxalic acid, to do exhaustive studies. The results showed that the conversion rate of hydrogen was changed from 3.03 to

3.95 mol/mol acetate when the voltage increased from 0.3 V to 0.8 V (Cheng et al., 2007). In addition, the input amount of the auxiliary voltage has effect on both the production efficiency and the hydrogen purity. Hydrogen content fell sharply and methane content were increased when the auxiliary voltage gradually reduced (Tartakovsky et al., 2008; Call et al., 2008). In 2007, Jenna Ditzig etc. treated domestic wastewater using graphite-granule packed-bed microbial reactors, and the results showed the $_{CE}$ was 26% in 0.41 V (Jenna et al., 2007).

3. EVALUATION OF THE PERFORMANCE OF MECs

Up to now, many indicators were used to assess the working situation of the MEC system, and they were as follows: hydrogen production efficiency, $_{CE}$, energy recovery, the removal rate of COD, etc. In addition, MEC system was used in the treatment of domestic sewage, industrial wastewater treatment and other aspects. For the reason, this system can remove both inorganic pollutants (nitrogen, phosphorus) and organic pollutants (nitrobenzene), which are just the focus in wastewater treatment. Furthermore at present the removal rate of COD is about 80% in the amplification technics (Cusick et al., 2011; Logan et al., 2010).

Recent studies have shown that sulfate and nitrate, as main inorganic pollutants, can be removed at the cathode of MECs (Coma et al., 2013; Zhan et al., 2012). Sulfate is an abundant pollutant found in groundwater. Recently, it has been reported that sulfate were removed as electron acceptor in a MEC, and the removal rate reached more than 50g $SO4^{2-}$ /m^3/d at 1.4 V (Coma et al., 2013). However, two major limitations are the low conductivity of groundwater and the large pH gradient between cathode and anode. About the nitrogen removal using MECs, the value was 92.6% at the voltage of 0.4 V. In this process, the DO level should be controlled because of the different demand for oxygen between nitrification and denitrification (Zhan et al., 2012). Generally, most of the organic pollutants can be removed using MECs such as nitrobenzene (NB), Chlorophenols (CPs). Moreover, the production rate of aniline reached to 3.06 mol/m^3/d, which is main product of NB after degradation (Mu et al., 2009a; Wang et al., 2012). CPs, as one of toxic and bio-refractory pollutants, its removal rate reached to 0.38 mol/m^3/d (Wen et al., 2012).

4. PROBLEMS AND CHALLENGES

Although hydrogen technology by MEC is still in the initial stage, some achievements have been done such as electrode materials and electrode distance, etc. But most of researches are limited in the labratories, there is a long way to the stage of industrialization (Jiang et al., 2012). Main reasons include as follows: (1) microbial hydrogen production efficiency was not high enough to meet the social demand for hydrogen, because of genetic and metabolic characteristics of the existing hydrogen-producing bacteria; (2) The accumulation of organic acids in anaerobic fermentation reactors lead to a lower pH, which restrict the normal growth and metabolism of microorganism, result in a decline of hydrogen production capacity, at the same time increase the cost from pH adjustment with alkaline, environmental pollution and other adverse effects; (3) The high cost of electrode materials and some related material, low power density, higher internal resistance; (4) Exoelectrogens is difficult to control in bigger reactors; (5) Former researches about hydrogen production by MEC mainly focus on project parameter design, conventional basic bacteria breeding, etc. Although great progress has been made, the current efficiency of hydrogen production is not sufficient to achieve industrial production level. At present, the researches about MEC mainly focus on some simple substrates, such as glucose, sodium acetate, propionate, domestic sewage etc., but there are few studies about some complex organic wastewater, further researches are needed for this. In addition, different organic loading have influence on hydrogen production of MECs under the same substrates. Escapa's research showed that glycerol was used as the substrate in the single chamber MEC, and the hydrogen efficiency increased with the increase of glycerol concentration, however, when the concentration of glycerol was over 2.7g/(La. d), hydrogen efficiency reduced with increasing of glycerol concentration. Additionally, overload phenomenon appeared in MEC, which was similar with He's study about double chamber MFC (Escapa et al., 2009; He et al., 2015). So it is imminent to further improve the efficiency of biological hydrogen production (Liu et al., 2008). In the near future, MEC system is expected to be used for the practical problem (Logan et al., 2012).

ACKNOWLEDGMENTS

This work was finally supported by Natural Science Foundation of China (21106080), the Chinese Universities Scientific Fund (2012RC030), the National Key Technology Support Program of China (2014BAD02B03), and NSFC-JST Cooperative Research Project (21161140328).

REFERENCES

[1]   Anders Thygesen, Massimo Marzorati, Nico Boon, Anne Belinda Thomsen, Willy Verstraete Upgrading of straw hydrolysate for production of hydrogen and phenols in a microbial electrolysis cell (MEC) [J]. *Applied microbiology and biotechnology*, 2011, 89: 855-865.

[2]   Bond D.R., Lovley D.R.. Electricity production by *Geobacter sulfurreducens* attached to electrodes [J]. *Applied and environmental microbiology,* 2003, (69):1548-1555.

[3]   Call D.F., Logan B.E.. Hydrogen production in a single chamber microbial electrolysis cell lacking a membrane [J]. *Environmental science & technology,* 2008, 42(9):3401-3406.

[4]   Call D.F., Logan B.E.. A method for high throughput bioelectrochemical research based on small scale microbial electrolysis cells [J]. *Biosensors and Bioelectronics*, 2011, 26: 4526-4531.

[5]   Chae K.J., Choi M.J., et al. Selective inhibition of methanogens for the improvement of biohydrogen production in microbial electrolysis cells [J]. *International journal of hydrogen energy*, 2010, 35: 13379-13386.

[6]   Chae K.J., Choi M.J., Lee J., et al. Biohydrogen production via biocatalyzed electrolysis in acetate-fed bioelectrochemical cells and microbial communityanalysis [J]. *International journal of hydrogen energy,* 2008, 33 (19): 5184-5192.

[7]   Cheng K.Y., Cord-Ruwisch R., Ho G.. Renewable energy for sustainable development in the Asia Pacific Region (eds. Jennings P, Ho G, Mathew K, Nayer C V). Melville, N. Y.: *American Institute of Physics*, 2007, 264-269.

[8]   Cheng S.A., Logan B.E.. Sustainable and efficient biohydrogen production via electrohydrogenesis [J]. *PNAS,* 2007, 104 (47): 18871-18873.

[9]   Cheng S.A., Logan B.E.. High hydrogen production rate of microbial electrolysis cell (MEC) with reduced electrode spacing [J]. *Bioresource Technology,* 2011, 102: 3571-3574.

[10] Cheng S.A, Logan B.E.. Sustainable and efficient biohydrogen production via electro-hydrogenesis [J]. *Proceedings of the national academy of sciences of the United States of America,* 2007, 104(47):18871-18873.

[11] Cheng S.A., Patrick Kiely, Logan B.E.. Pre-acclimation of a wastewater inoculum to cellulose in an aqueous-cathode MEC improves power generation in air-cathode MFCs [J]. *Bioresource technology,* 2011, 102: 367-371.

[12] Clauwaert P., Toledo R., Vander Ha D., et al. Combining biocatalyzed electrolysis with anaerobic digestion[J]. *Water science and technology,* 2008.57(4):575-579.

[13] Coma M., Puig S., Pous N., et al. Biocatalysed sulfate removal in a BES cathode [J]. *Bioresource technology,* 2013, 130: 218-223.

[14] Cusick R.D., Bryan B., Parker D., et al. Performance of a pilot-scale continuous flow microbialelectrolysis cell fed winery wastewater [J]. *Applied microbiology and biotechnology,* 2011, 89: 2053-2063.

[15] Derek R., Lovle. Anaerobic production of magnetite by a dissimilatory iron-reducing microorganism [J]. *Nature,* 1987, 330 (6145): 252-254.

[16] Douglas F.C., Logan B.E.. Hydrogen production in a single chamber microbial electrolysis cell lacking a membrane [J]. *Environmental science & technology,* 2008, 42(9): 3401-3406.

[17] Douglas F.C., Logan B.E.. A method for high throughput bioelectrochemical research based on small scale microbial electrolysis cells. *Biosensors and bioelectronics,* 2011, 26: 4526-4531.

[18] Douglas F.C., Rachel C. Wagner, B.E. Logan. Hydrogen production by *Geobacter Species* and a mixed consortium in a microbial electrolysis cell [J]. *Applied and environmental microbiology,* 2009, 75(24): 7579-7587.

[19] Edgar Ribot-Llobet, Joo-Youn Nam, Justin C. Tokash, Albert Guisasola, Bruce E. Logan. Assessment of four different cathode materials at different initial pHs using unbuffered catholytes in microbial electrolysis cells [J]. *International journal of hydrogen energy,* 2013, 38: 2951-2956.

[20] Elizabeth S. Heidrich, Stephen R. Edwards, Jan Dolfing, Sarah E. Cotterill, Thomas P. Curtis. Performance of a pilot scale microbial electrolysis cell fed on domestic wastewater at ambient temperatures for a 12 month period [J]. *Bioresource technology,* 2014, 173: 87-95.

[21] Elsemiek Croese, Maria Alcina Pereira, Gert-Jan W Euverink. Analysis of the microbial community of the biocathode of a hydrogen producing microbial electrolysis cell [J]. *Applied microbiology and biotechnology,* 2011, 92:1083-1093.

[22] Escapa A., Gil-Carrera L., García V., Morán A.. Performance of a continuous flow microbial electrolysis cell (MEC) fed with domestic wastewater [J]. *Bioresource technology*, 2012, 117: 55-62.

[23] Escapa A., Manuel M.F., Moran A., et al. Hydrogen Production from Glycerol in a Membraneless Microbial Electrolysis Cell [J]. *Energy fuels*, 2009, 23: 4612-4618.

[24] Jack R. Ambler, Logan B.E.. Evaluation of stainless steel cathodes and a bicarbonate buffer for hydrogen production in microbial electrolysis cells using a new method for measuring gas production [J]. *International journal of hydrogen energy*, 2011, 36: 160-166.

[25] Jenna Ditzig, Liu Hong, Logan B.E. Production of hydrogen from domestic wastewater using a bioelectrochemically assisted microbial reactor (BEAMR) [J], *International journal of hydrogen energy*, 2007, 32: 2296-2304.

[26] Jeremiasse A.W., Hamelers E.V.M., Buisman C.J.N.. Microbial electrolysis cell with a microbial biocathode [J]. *Bioelectrochemistry*, 2010, 78(1): 39-43.

[27] Jiang Y.Y., Xu Y., Chen Y.W., et al. Progress in hydrogen production using microbial electrolysis cell [J]. *Modern chemical industry*, 2012, 10:34-38.

[28] Jooyoun N., Justin C. Tokash, Logan B.E.. Comparison of microbial electrolysis cells operated with added voltage or by setting the anode potential [J]. *International journal of hydrogen energy*, 2011, 36: 10550-10556.

[29] Hawkes F.R., Dinsdale R., Hawkes D.L., et al. Sustainable fermentative hydrogen production: challenges for process optimization [J]. *International journal of hydrogen energy*, 2002, 27 (11/12): 1339-1347.

[30] He Y.H., Liu Z.D., et al. Carbon nanotubes simultaneously as the anode and microbial carrier for up-flow fixed-bed microbial fuel cell [J]. *Biochemical engineering journal*, 2015, (94):39-44.

[31] Hu H.Q., Fan Y.Z., Liu H. Hydrogen production using single-chamber membrane-free microbial electrolysis cells [J]. *Water research*, 2008, 42: 4172-4178.

[32] Huang X., Liang P., Cao X.X., Fan M.Z.. Progress in research of mediator-less microbial fuel cells [J]. *China water & wastewater*, 2007, 23(4):1-6.

[33] Guo K., Zhang J.J., Li H.R., Du Z.W. Hydrogen production by microbial electrolysis cells [J]. *Progress in chemistry*, 2010, 22(4):748-753.

[34] Kim H.J., Park H.S., Hyun M.S., Chang I.S., Kim M., Kim B.H.. A mediator-less microbial fuel cell using a metal reducing bacterium, Shewanella putrefaciens [J]. *Enzyme and microbial technology*, 2002, (30): 145-152.

[35] Kim I.S., Hwang M.H., Jang N.J., et al. Effect of low pH on the activity of hydrogen utilizing methanogen in bio-hydrogen process [J]. *International journal of hydrogen energy*, 2004, 29(11): 1133-1140.

[36] Lee H.S., Torres C.I., Parameswaran P., et al. Fate of H2 in an upflow single-chamber microbial electrolysis cell using a metal-catalyst-free cathode [J]. *Environmental science & technology*, 2009, 43(20): 7971-7976.

[37] Liu, H., Cheng S., Logan B.E. Production of electricity from acetate or butyrate in a single chamber microbial fuel cell [J]. *Environmental science & technology*, 2005, (39):658-662.

[38] Liu H., Grot S., Logan B.E.. Electrochemically assisted microbial production of hydrogen from acetate [J]. *Environmental science & technology*, 2005, 39(11): 4317-4320.

[39] Liu, H., Logan B.E. Electricity generation using an air-cathode single chamber microbial fuel cell in the presence and absence of a proton exchange membrane [J]. *Environmental science & technology*, 2004, 38: 4040-4046.

[40] Liu W.Z.. Hydrogen generation from organic wastewater in microbial electrolysis cells and function analysis of anodophilic communities [D]. *Harbin institute of technology*, 2011, 6.

[41] Liu W.Z., Huang S.C., Zhou A.J., et al. Hydrogen generation in microbial electrolysis cell feeding with fermentation liquid of waste activated sludge [J]. *International journal of hydrogen energy*, 2012, 37: 13859-13864.

[42] Liu W.Z., Wang A.J., Sun D., Ren N.Q., et al. Characterization of microbial communities during anode biofilm reformation in a two-chambered microbial electrolysis cell (MEC) [J]. *Journal of biotechnology*, 2012, 157: 628-632.

[43] Liu X.M., Ren N.Q., Song F.N. Recent advances in biohydrogen production by microbe fermentation [J]. *Acta energiae solaris sinica*, 2008, 29(5): 544-549.

[44] Logan B.E., Call D., Cheng S., Hamelers H.V.M., Sleutels T.H.J.A., Jeremiasse A.W., Rozendal R.A.. Microbial electrolysis cells for high yield hydrogen gas production from organic matter [J]. *Environmental science & technology*, 2008, 42(23): 8630-8640.

[45] Logan B.E., Douglas C., Cheng S.A., et al. Microbial electrolysis cells for high yield hydrogen gas production from organic matter [J]. *Environmental science & technology*, 2008, 42, (23): 8630-8640.

[46] Logan B.E., Hamelers B., Rozendal R., et al. Microbial fuel cells: methodology and technology [J]. *Environmental science & technology*, 2006, 40(17): 5181-5192.

[47] Logan B.E., Rabaey K. Conversion of wastes into bioelectricity and chemicals by using microbial electrochemical technologies [J]. *Science*, 2012, 337(6095): 686-690.

[48] Logan B.E.. Scaling up microbial fuel cells and other bioelectrochemical systems [J]. *Applied microbiology and biotechnology*, 2010, 85(6): 1665-1671.

[49] Lu L., Ren N.Q., Zhao X., et al. Hydrogen production, methanogen inhibition and microbial community structures in psychrophilic single-chamber microbial electrolysis cells [J]. *Energy & environmental science*, 2011.

[50] Lu L., Xing D.F., Ren N.Q., Logan B.E.. Syntrophic interactions drive the hydrogen production from glucose at low temperature in microbial electrolysis cells [J]. *Bioresource technology*, 2012, 124: 68-76.

[51] Lu L., Xing D.F., Xie T.H., Ren N.Q., Logan B.E.. Hydrogen production from proteins via electrohydrogenesis in microbial electrolysis cells [J]. *Biosensors and bioelectronics*, 2010, 25: 2690-2695.

[52] Lu Y., Zhang C., Xing X.H.. Effect of cultivation conditions on hydrogen production by *Clostridium paraputrificum* M-21 [J]. *Chinese journal of bioprocess engineering*, 2004, 2(2): 41-45.

[53] Luo H.P., Fu S.Y., Liu G.L. et al. Autotrophic biocathode for high efficient sulfate reduction in microbial electrolysis cells [J]. *Bioresource technology*, 2014, 167: 462-468.

[54] Min B., Cheng S., Logan B.E... Electricity generation using membrane and salt bridge microbial fuel cells [J]. *Water research*, 2005, (39):1675-1686.

[55] Mu, Y., Rozendal, R.A., Rabaey, K., Keller, J.. Nitrobenzene removal in bioelectrochemical systems [J]. *Environmental science & technology*, 2009a, 43(22): 8690-8695.

[56] Nath K., Das D.. Improvement of fermentative hydrogen production: various approaches [J]. *Applied microbiology and biotechnology*, 2004, 65: 520-529.

[57] Nuria Montpart, Laura Rago, Juan A. Baeza, Albert Guisasola. Hydrogen production in single chamber microbial electrolysis cells with different complex substrates [J]. *Water research*, 2015, 68: 601-615.

[58] Oh S.E., Van Ginkel S, Logan B.E.. The relative effectiveness of pH control and heat treatment for enhancing biohydrogen gas production [J]. *Environmental science &. technology*, 2003, 37(22):5186-5190.

[59] Ren L.J., Michael Siegert, Ivan Ivanov, John M. Pisciotta, Bruce E. Logan. Treatability studies on different refinery wastewater samples using high-throughput microbial electrolysis cells (MECs) [J]. *Bioresource technology*, 2013, 136: 322-328.

[60] Roland D. Cusick, Bill Bryan, Denny S. Parker, Matthew D. Merrill, Maha Mehanna, Patrick D. Kiely, Guangli Liu, Bruce E. Logan. Performance of a pilot-scale continuous flow microbial electrolysis cell fed winery wastewater [J]. *Applied microbiology and biotechnology,* 2011, 89: 2053-2063.

[61] Roland D. Cusick, Logan B.E.. Phosphate recovery as struvite within a single chamber microbial electrolysis cell [J]. *Bioresource technology*, 2012, 107: 110-115.

[62] Rozendal R.A., Hamelers H.V.M., Euverink G.J.W., et al. Principle and perspectives of hydrogen production through biocatalyzed electrolysis [J]. *International journal of hydrogen Energy*, 2006, 31(12): 1632-1640.

[63] Rozendal R.A., Hamelers H.V.M., Rabaey K., et al. Towards practical Implementation of bioelectrochemical wastewater treatment [J]. *Trends in biotechnology*, 2008, 26:450-459.

[64] Rozendal R.A., Jeremiasse A.W., Hamelers H.V.M., et al. Hydrogen production with a microbial biocathode[J]. *Environmental science & technology*, 2008, 42(2): 629-634.

[65] Selembo P.A., Merrill M.D., Logan B.E.. The use of stainless steel and nickel alloys as low-cost cathodes in microbial electrolysis cells [J]. *Journal of power sources,* 2009, 190(2): 271-278.

[66] Sun M., Sheng G.P., Zhang L., et al. An MEC-MFC-Coupled system for biohydrogen production from acetate [J]. *Environmental science & technology,* 2008, 42: 8095-8100.

[67] Tanisho S., Kuromoto M.. Effect of CO_2 removal on hydrogen production by fermentation [J]. *INTJ hydrogen energy*, 1998, 23(7):559-563.

[68] Tartakovsky B., Manuel M.F., Neburchilov V., et al. Biocatalyzed hydrogen production in a continuous flow microbial fuel cell with a gas phase cathode [J]. *Journal of power sources*, 2008, 182(1):291-297.

[69] Wang A.J., Cui D., Cheng H.Y., et al. A membrane-free, continuously feeding, single chamber up-flow biocatalyzed electrolysis reactor for nitrobenzene reduction [J]. *Journal of hazardous materials*, 2012, 199: 401-409.

[70] Wang Z.J., Zheng Y., Xiao Y. et al. Analysis of oxygen reduction and microbial community of air-diffusion biocathode in microbial fuel cells [J]. *Bioresource technology*, 2013, 144:74-79.

[71] Wen, Q., Yang, T., Wang, S., Chen, Y., Cong, L., Qu, Y. Dechlorination of 4-chlorophenol to phenol in bioelectrochemical systems [J]. *Journal of hazardous materials*, 2012, 244: 743-749.

[72] Xing X.H., Zhang C.. Research progress in dark microbial fermentation for bio-hydrogen production [J]. *Chinese journal of bioprocess engineering*, 2005, 3(1): 1-8.

[73] Yokoi H., Maki R., et al. Microbial production of hydrogen from starch-manufacturing wastes [J]. *Biomass and bioenergy*, 2002, 22:385-395.

[74] Zhan, G., Zhang, L., Li, D., Su, W., Tao, Y., Qian, J. Autotrophic nitrogen removal from ammonium at low applied voltage in a single-compartment microbial electrolysis cell [J]. *Bioresource technology*, 2012, 116: 271-277.

[75] Zhang Y.F.. Angelidaki Irini. Microbial electrolysis cells turning to be versatile technology: Recent advances and future challenges [J].*Water research*, 2014, 56: 11-25.

In: Gas Biofuels from Waste Biomass
Editor: Zhidan Liu

ISBN: 978-1-63483-192-5
© 2015 Nova Science Publishers, Inc.

Chapter 8

TWO-STAGE ANAEROBIC FERMENTATION FOR CO-PRODUCTION OF HYDROGEN AND METHANE

Xi-Yu Cheng[*]

College of Life Sciences and Bioengineering, School of Science,
Beijing Jiaotong University, Beijing, China

ABSTRACT

Anaerobic digestion processes have been widely applied for methane or hydrogen production from waste biomass. Recently, a remarkable two-stage dark fermentation process for co-production of hydrogen and methane has drawn much attention due to its high efficiency in substrate utilization and energy recovery. In this chapter, an up-to-date overview of current knowledge about co-production of hydrogen and methane via a two-stage dark fermentation process from various biomass wastes is presented. This chapter mainly includes four parts: an overview of waste biomass, co-production of hydrogen and methane from the waste biomass in the two-stage anaerobic fermentation, a comparison of the two-stage dark fermentation with other hybrid processes and concluding remarks.

Keywords: dark fermentation, waste biomass, hybrid process, hydrogen, methane

ABBREVIATIONS

ASBR, anaerobic sequencing batch reactor; CSTR, continuous stirred tank reactor; COD, chemical oxygen demand; DMC, direct microbial conversion; HR: hydrogen-producing reactor; HRT, hydraulic retention time; HAB, homoacetogenic bacteria; LCFA, long chain fatty acids; MPB, methane-producing bacteria; MR: methane-producing reactor; OLR:organic

[*]Corresponding Author address: College of Life Sciences and Bioengineering, School of Science, Beijing Jiaotong University, Beijing 100044, P.R. China. Tel.: +86-10-51684351-209, Fax: +86-10-51683887, Email:xycheng@bjtu.edu.cn

loading rate; SRB, sulfatereducing bacteria; UASB, upflow anaerobic sludge blanket; VS, volatile solids; VFA, volatile fatty acids.

INTRODUCTION

With rising energy demands, environmental pollution, and concerns about looming climatic changes, interest in production of renewable energy, such as hydro, wind, solar and biomass energy, has grown (Cheng and Zhong, 2014; Demirel et al., 2010; Li et al., 2014; Li et al., 2015; Li and Liu, 2014; Liu et al., 2011; Liu et al., 2012; Park et al., 2010; Rabelo et al., 2011; Shen and Luo, 2015). Conversion of biomass to energy can be carried out by different processes, including thermochemical processes, such as combustion (heat/electricity) and gasification (syngas), physicochemical processes (biodiesel), or biochemical processes (bio-ethanol, bio-butanol or bio-methane and bio-hydrogen) (Nissila et al., 2014; Srirangan et al., 2012). The main advantages of biomass energy are the local availability of biomass, its renewability, feasibility of waste biomass to energy, low requirements of capital investments, and reduction of greenhouse gas emissions (Hoogwijk et al., 2003; Nissila et al., 2014).

Hydrogen, which can be used directly in combustion engines for transportation or in fuel cells for producing electricity, is widely considered as an ideal energy carrier due to its advantages, such as a high efficiency of conversion to usable power, a high gravimetric energy density and the fact that it is environmentally-friendly (Nissila et al., 2014; Elsharnouby et al., 2013). Currently, most of commercial hydrogen (88%) derives from fossil fuels such as natural gas, heavy oils or coal (Guo et al., 2010a; Nath and Das, 2003), while water electrolysis has extensively developed in recent years and supplies up to 4% of hydrogen production. However, all of these processes are energy-extensive and unsustainable. Biological hydrogen production processes, which include direct water biophotolysis by green algae, indirect water biophotolysis by cyanobacteria, the photo fermentation by photosynthetic bacteria, and dark fermentation by strict or facultative anaerobic bacteria, provide one promising alternative because of less energy requirement. Dark hydrogen fermentation, in which hydrogen is able to be produced with a high rate from various organic wastes, is regarded as one of the most attractive avenues (Levin and Chahine, 2010). In the last decades, successful hydrogen production from a number of potentially suitable waste biomass, such as crop residues, manure wastes, organic fractions of municipal solid waste (OFMSW), food waste, and algae, was carried out using dark fermentation. Lee and Chung (2010) reported that dark fermentation of food waste may be economically viable. However, it is also generally acknowledged that the majority of the organic content of the original substrate remains un-degraded after hydrogen fermentation. Due that energy recovery rate of single anaerobic hydrogen fermentation is about 3-10%, it is still-difficult for industrial application at present (Cheng et al., 2012a; Levin et al., 2007; Lu et al., 2009). Co-production of hydrogen and methane in two-stage dark fermentation, with advantages of simple operation, high production rate and wide substrate streams, was regarded as a potential avenue for improving energy recovery (Lu et al., 2009; Park et al., 2010).

This chapter presents an up-to-date overview of current knowledge about co-production of hydrogen and methane via a two-stage dark fermentation process from various waste biomass. In the chapter, characteristics of waste biomass and co-production of hydrogen and methane from different kinds of waste biomass will be summarized. Then, the effect of process conditions on the two-stage dark fermentation (mainly the key hydrogen fermentation) will be further discussed. Finally, comparison of the two-stage dark fermentation with other hybrid processes as well as areas where further attention is required will be presented.

OVERVIEW OF WASTE BIOMASS

Waste Biomass Resource

A wide spectrum of carbohydrates can be used for bio-hydrogen production. Although nearly 80% of the studies reported in the surveyed literature have focused on fermentation of pure sugars, such as glucose and sucrose (Elsharnouby et al., 2013), a number of other researchers present successful conversion of biomass to bio-hydrogen via dark fermentation. Given the renewable nature of biomass, these results provide a promising alternative for sustainable energy production. The main waste biomass sources include agricultural crops and their waste byproducts, wood wastes, OFMSW, animal wastes, food wastes, and aquatic plants and algae, etc. (Demirbas et al., 2009; Kirtay et al., 2011). Biomass is the name given all the earth's living matter. Biomass as the solar energy stored in chemical form in plant and animal materials is among the most precious and versatile resources on earth. Biomass is produced by green plants converting CO_2 in the air, water and sunlight into plant material through photosynthesis (Tuncel et al., 2004). Photosynthesis is a carbon fixation reaction by reduction of CO_2. The fixation or reduction of CO_2 is a light-independent process. It is composed mainly of carbohydrate compounds, in which the major elements are carbon, hydrogen, oxygen and nitrogen. Lignocellulosic biomass, a main part of waste biomass, is regarded as one of the most attractive feedstocks for renewable energy production (Levin and Chahine, 2010). The annual, worldwide production of lignocellulosic biomass reach about 220 Pg (dry weight) mainly consisting of agricultural, forestry and food processing residues, aquatic plants, energy crops, and algae (Chandra et al., 2012; Kirtay et al., 2011).

Components of Biomass

The components of biomass include cellulose, hemicelluloses, lignin, lipids, proteins, extractives, mono-sugars, starches, hydrocarbons, ash, water, and other compounds. Cellulose and hemicelluloses are two large carbohydrate categories, while the lignin fraction consists of non-sugar type molecules (Demirbas, 2009).

The components of biomass have been well reviewed (Hendriks and Zeeman, 2009; Mosier et al., 2005; Saratale et al., 2008). Lignocellulosic biomass consists of mainly cellulose, hemicelluloses and lignin, which are associated with each other (Fengel and Wegener, 1984). In brief, cellulose exists of D-glucose subunits, links by β-1, 4 glycosidic bonds. The cellulose in biomass consists of parts with a crystalline structure, and parts with a amorphous structure. Hemicelluloses, which exhibit lower molecular weights than cellulose, is a mixture of various polymerized monosaccharides such as glucose, xylose, mannose, galactose, arabinose, and sugar acids. The dominant component of hemicelluloses from agricultural biomass and hardwood is xylan, while this is glucomannan for softwood. The number of repeating saccharide monomers is only 150, against 5000–10000 in cellulose (Saratale et al., 2008). Lignin, which gives the plant structural support, impermeability, and resistance against attack of microorganisms and oxidative stress, is one of the most abundant polymers in nature and is present in the cellular wall (Hendriks and Zeeman, 2009). Lignin, an amorphous heteropolymer, consists of three different phenylpropane units (primarily syringyl, guaiacyl and p-hydroxy phenol), that are combined together by various kind of linkages to form a very complex matrix (Balat et al., 2009).

Hemicelluloses serve as a connection between the cellulose fibers and the lignin, which makes the whole cellulose-hemicellulose-lignin network more rigidity (Laureano-Perez et al., 2005). Significant decrease of overall-degradation efficiency occurred with lignocellulosic biomass due to lignin physical barrier (Magnusson et al., 2008). All this has therefore necessitated pretreatment steps for cell structure destruction and delignification prior to hydrogen fermentation (Rani et al., 1998).

CO-PRODUCTION OF HYDROGEN AND METHANE FROM WASTE BIOMASS IN TWO-STAGE DARK FERMENTATION SYSTEM

While many studies investigated hydrogen production by dark fermentation of simple sugars, increasing research attentions have looked into conversion of the waste biomass. Due that energy recovery of single anaerobic hydrogen fermentation is about 3-10%, it is difficult for industrial application at present (Cheng et al., 2012a; Levin et al., 2007; Lu et al., 2009). Co-production of hydrogen and methane in the two-stage dark fermentation, with advantages of simple operation, high production rate and wide substrate streams, was regarded as a potential avenue for improving energy recovery (Lu et al., 2009). Another important advantage of the two-stage dark fermentation system is that it is possible to produce hydrogen during the first acidogenic phase, and subsequently to produce methane during the second phase. In this section, co-production of hydrogen and methane from biomass in the two-stage dark fermentation system are exclusive discussed from the following aspects:

❖ Hydrogen and methane potential of different biomass
❖ Effect of process conditions
❖ Energy recovery of two-stage system
❖ Pilot scale studies of two-stage system

Hydrogen and Methane Potential of Different Biomass

Research activities on co-production of hydrogen and methane mainly focused on four kinds of waste biomass, *i.e.,* crop residues, organic fractions of municipal solid waste& food waste, livestock waste (animal manure), and algae biomass.

Crop Residue

Agricultural residues from harvested crops, which include stalks, stover, peelings, cobs, straw, bagasse, and other lignocellulosic residues, are the most abundant, cheapest and most readily available organic waste source to be biologically converted (Muti, 2009). The annual yield of lignocellulosic biomass from the primary agricultural sector has been evaluated at approximately 200 billion tons worldwide (Ren et al., 2009). All agricultural crop residues are biodegradable and, to varying degrees, may be converted biologically in dark fermentation processes to hydrogen and methane.

Hydrogen yields in dark fermentation of various crop substrates, as extensively recorded in the literature. Untreated raw material indicates generally lower yields in one stage dark hydrogen fermentation, ranging from 0.5 to 16 ml H_2 g/VS (Guo et al., 2010a). Increased biogas (including hydrogen and methane) yields and energy recovery were observed in the two-stage dark fermentation process (Table 1). Kim et al., (2013) evaluated co-production of hydrogen and methane in a two-stage batch thermophilic fermentation with untreated sewage sludge (SS) and rice straw (RS) as raw materials. In the first stage H_2 fermentation process using untreated RS with raw SS, a high H_2 yield (21 ml/g-VS) and stable H_2 content (60.9%) was obtained. Direct utilization of post-H_2 fermentation residues readily produced biogas, and significantly enhanced the CH_4 yield (266 ml/g-VS) with stable CH_4 content (75~80%) during the second stage. Overall, volatile solids removal (60.4%) for the two-stage system was 37.9% higher than that in one-stage methane fermentation system. The efficient production of bioenergy is believed to be due to a synergistically improved second stage process exploiting the well-digested post-H_2 generation residues over the one-stage system.

A comparable biogas yields were observed in a two-stage fermentation of high concentration of corn stalk (Guo et al., 2014). The maximum hydrogen yield of 79.8 ml/g-TS and hydrogen production rate of 3.78 ml/g-cornstalk h was observed at fixed acidizing-cornstalk of 60 g/l, strains *Bacillus* sp. F 2011 dosage of 10%(v/v), CTAB of 30 mg/l, NH_4HCO_3 of 1.2 g/l and initial pH of 7.5 at 36 $^{\circ}$C, respectively. The hydrogen fermentation residues were further employed as the feedstock to produce methane by methanogenic bacteria and the maximum methane yield of 227 ml CH_4/g-COD and COD removal rate of 95% were recorded at 55°C.

The hydrogen yield in dark fermentation of crop residues in thermophilic conditions was higher than that in mesophilic conditions, which indicates that high temperature favors hydrolysis (Karlsson et al., 2008). The two-stage process for extreme thermophilic hydrogen and thermophilic methane production from wheat straw hydrolysate was investigated in up-flow anaerobic sludge bed (UASB) reactors (Kongjan et al., 2011). Specific hydrogen and methane yields of 89 ml/g-VS (190 ml/g-sugars) and 307 ml CH_4/g-VS, respectively were achieved simultaneously with the overall VS removal efficiency of 81% by operating with total hydraulic retention time (HRT) of 4 days. Dominant hydrogen-producing bacteria in the H_2-UASB reactor were *Thermoanaerobacter wiegelii, Caldanaerobacter subteraneus,* and *Caloramator fervidus*. Meanwhile, the CH_4-UASB reactor was dominated with methanogens of *Methanosarcina mazei* and *Methanothermobacter defluvii*.

Table 1. Co-production of hydrogen and methane in two-stage dark fermentation of waste biomass

Substrate	Inoculum of HR	Reactor operation	H_2 production	CH_4 production	Energy recovery (% or kJ/g TS)	Reference
Cornstalk	Heat-treated sludge and *C. thermocellum*	HR: 70 ml batch reactor at pH 6 and 55 °C; MR: 70 ml batch reactor at pH 7.5 and 37°C	63.7 ml/g-CS	114.6 ml/g-CS	54.1% (4.8 kJ/g TS)	Lu et al., 2008
Cornstalk	*C. thermocellum* DSM 7072	HR: 125 ml batch reactor at pH 7.2 and 55 °C (Control: one stage hydrogen fermentation)	37.6 ml/g-CS	--	3.7% (0.44 kJ/g TS)	Cheng and Liu, 2012a
Cornstalk	*C. thermocellum* DSM 7072	HR: 125 ml batch reactor at pH 7.2 and 55 °C; MR: 125 ml batch reactor at pH 7.2 and 35 °C	74.4 ml/g-CS	205.8 ml/g-CS	70.0% (8.3 kJ/g TS)	Cheng et al., 2012
Cornstalk	*C. thermocellum* DSM 7072	HR: 10 l batch CSTR at pH 7.2 and 55 °C; MR: 10 l continuous UASB at pH 7.2, 35 °C, and HRT of 10 h	58.0 ml/g-CS	200.9 ml/g-CS (4 l/l-d)	67.2% (7.9 kJ/g TS)	Guo et al., 2014
Cornstalk	Cow dung compost after HST	HR: 4 l batch reactor at pH 7.5 and 36 °C; MR: 6.5 l batch reactor at pH 7.2 and 55 °C	79.8 ml/g-CS	227 ml/g-COD	--	
Algae biomass	--	MR: 500 ml batch reactor at pH 7.0 and 55 °C with addition of *C. thermocellum* DSM 2036 (Control: one stage methane fermentation)	--	316.6 ml/g-VS	12.2 kJ/g VS	Lü et al., 2013
Algae biomass	*C. thermocellum* DSM 2036	HR: 500 ml batch reactor at pH 5.5 and 55 °C; MR: 500 ml batch reactor at pH 7.0 and 55 °C	53.4 ml/g-VS$_{fed}$	320.6 ml/g-VS$_{fed}$	13.4 kJ/g VS	
Household solid waste	Sludge acclimated at a short HRT	HR: 1 l continuous CSTR at pH 5, 37 °C, OLR of 37.5 g-VS/l-d and SRT of 2 d; MR: 4.5 l continuous CSTR at pH 7.5, 37 °C, OLR of 4.1g-VS/l-d and SRT of 15 d	43 ml/g-VS$_{fed}$	500 ml/g-VS$_{fed}$	--	Liu et al., 2006
Activated sludge and food waste (VS, 15:85)	Activated sludge after HST at 100 °C for 30 min	HR: 300 ml batch reactor at pH 5.5 and 37°C; MR: 300 ml batch reactor at pH 7.0 and 37°C	106.4 ml/g-VS$_{fed}$	353.5 ml/g-VS$_{fed}$	14.0 kJ/g-VS	Liu et al., 2013
Cow manure and waste milk (VS, 30:70)	Manure after HST	HR: 1 l batch reactor at pH 6.2 and 55°C; MR: 1 l batch reactor at pH 7.5 and 55°C	38.2 ml/g-VS	627.6 ml/g-VS	25.5 kJ/g-VS	Lateef et al., 2014
Sewage sludge and rice straw	Sewage sludge	HR: 500 ml batch reactor at pH 7.0 and 55°C; MR: 500 ml batch reactor at pH 7.2 and55°C	21 ml/g-VS (H_2%: 60.9%)	266 ml/g-VS (CH_4%: ~80%)	8.8 kJ/g-VS	Kim et al., 2013
Biowaste	Sludge acclimated at a short HRT	HR: 200 l continuous CSTR at pH 5.5, 55 °C, OLR of 16 kgTVS/m³d and HRT of 3.3 d; MR: 760 l continuous CSTR at pH 8.2, 55 °C, OLR of 4 kgTVS/m³ d and HRT of 12.6 d	51 ml/g-VS$_{fed}$ (H_2%: 37%) (2.3 l/l-d)	416 ml/g-VS$_{fed}$ (CH_4%: 65%) (2.7 l/l-d)	--	Cavinato et al., 2011
Food waste	Activated sludge after HST at 80 °C for 20 min	HR: 500 l continuous CSTR at pH 5.5, 33 °C, OLR of 12-71 gCOD/l and HRT of 21 h; MR: 2300 l continuous UASB at pH 7.4, 36 °C, OLR of 2.7-6.4 gCOD/l-d and HRT of 3.9 d	1.8 mol H_2/mol hexose 3.9 l/m³d (H_2%: 60%)	6 l/l-d (CH_4%.81-83%)	--	Lee and Chung, 2010

Note: HR: hydrogen-producing reactor; MR: methane-producing reactor; Ref.: Reference; CS: cornstalk; *C. thermocellum*: *Clostridium thermocellum*; BESA:.2-bromoethanesulfonic acid.

The composition of different crops residues, e.g., wheat straw, corn stover and rice straw, contain cellulose, hemicelluloses and lignin in a range of approx. 32~47%, 19~27% and 5~24%, respectively (Saratale et al., 2008).

It is well known that the hydrolysis is regarded as a rate-limiting step in the hydrogen fermentation of lignocellulosic biomass due to its recalcitrant structure. Therefore, appreciate pretreatment steps for the raw materials are often required in order to improve accessibility of such raw materials. The hydrogen yield from cornstalks treated by NaOH (0.5%) reached 57 ml/g-VS, against 3 ml/g-VS observed in the case of untreated cornstalks (Zhang et al., 2007). Steam explosion, thermo-acid pretreatment and microwave-assisted acid/alkali pretreatment were also proved as useful methods to improve the efficiency of substrate degradation and the following fermentative hydrogen production (Cheng and Liu, 2010; Cheng et al., 2012b; Datar et al., 2007; Liu and Cheng, 2009, 2010; Xing et al., 2010).

It should be noted that most of chemical and thermo pretreatments are energy extensive processes and require specific equipment and/or vessels. The use of acid/alkali may produce additional environmental pollution. Therefore, researchers have tried to develop biological strategies to improve the hydrolysis of lignocellulosic biomass. External cellulase addition, which is proved as an efficient method for the enhancement of hydrogen fermentation (Li and Chen, 2007), may not be an economically feasible one till now because of high cost of cellulase production. Alternatively, thermophilic hydrogen fermentation enhanced by a fungal pretreatment process, in which raw cornstalk was fermented together with the pretreated cornstalk by cellulase-producing fungi, *Trichoderma reesei* Rut C-30, provides a relatively promising avenue for bioconversion of waste biomass to renewable hydrogen energy (Cheng and Liu, 2012b).

Recently, co-production of hydrogen and methane based on direct microbial conversion (DMC) with thermophilic baterica with a strong cellulose-degrading ability, e.g., *Clostridium thermocellum*, attracted increasing attention. A three-stage anaerobic fermentation process including H_2 fermentation I, H_2 fermentation II and methane fermentation was developed for the co-production of hydrogen and methane from cornstalks. Hydrogen production from cornstalks using DMC by *Clostridium thermocellum* DSM 7072 was markedly enhanced in the two-stage thermophilic hydrogen fermentation process integrated with alkaline treatment. The highest total hydrogen yield from cornstalks in the two-stage fermentation process reached 74.4 ml/g-cornstalk. The hydrogen fermentation effluents and alkaline hydrolyzate were further used for methane fermentation by anaerobic granular sludge, and the total methane yield reached 205.8 ml/g-cornstalk (Cheng and Liu, 2012a). This process has been successfully scaled up from 125 ml anaerobic bottles to lab scale bioreactors (Cheng et al., 2012). A 10 l continuous stirred tank reactor (CSTR) system was used for hydrogen fermentation. The maximum hydrogen production rate reached 218.5 ml/l h at a cornstalk concentration of 30 g/l, and the total hydrogen yield and volumetric hydrogen production rate reached 58.0 ml/g-cornstalk and 0.55–0.57 l/l-d, respectively. The effluents of hydrogen fermentation were continuously digested for methane fermentation in a 10 l up-flow anaerobic sludge bed (UASB). At an organic loading rate of 15.0 g-COD/l-d, the COD removal efficiency and volumetric biogas production rate reached 83.3% and 4.6 l/l-d, respectively.

Food Waste and Organic Fractions of Municipal Solid Waste

A number of potentially suitable residual substrates have been investigated for biohydrogen production by dark fermentation. Among these, fractions of municipal solid waste such as food waste and the broader mixture of components known as organic fractions of municipal solid waste (OFMSW; basically food waste combined with non-recoverable paper residues) may represent relatively inexpensive and suitable substrates for hydrogen fermentation, mainly because of their high carbohydrate content and wide availability (Okamoto et al., 2000; Lay et al., 2003; Kim et al., 2011a; Zhu et al., 2008; Wang and Zhao, 2009). Food waste, which has high energy content and is highly biodegradable, contains 85~95% of volatile solids and 75~85% moisture and favors microbial-development (Li et al., 2008a). Over the past decades, OFMSW/food waste has been the most studied feedstock for hydrogen production, including kitchen refuse (Jayalakshmi et al., 2009), food industry co-products such as oil mill (Eroglu et al., 2009; O-Thong et al., 2008a), and starch-manufacturing waste (Yokoi et al., 2002).

The performances of hydrogen fermentation of OFMSW display great variations, from 3 ml/g-VS to more than 290 ml/g-VS (Guo et al., 2010a; Karlsson et al., 2008; Koutrouli et al., 2009; Okamoto et al., 2000; Wang et al., 2009), which is because of the different composition of the matter involved. The average production is substantially higher than the values observed from fermentation of crop residues and livestock. A two-stage process combined hydrogen and methane production from household solid waste (mainly food residue, paper, garden waste (leaves, branches)) has been successfully developed (Liu et al., 2006). The yield of 43 ml/g-VS_{added} was generated in the first hydrogen production stage and the methane production in the second stage was 500 ml CH_4/g-VS_{added}, which was 21% higher than the methane yield from the one-stage process. Sparging of the hydrogen reactor with methane gas resulted in doubling of the hydrogen production.

Food waste and OFMSW have often been considered for co-digestion with other organic wastes, including activated sludge and pulp & paper sludge as well as agricultural wastes. The use of co-substrates is motivated by the following considerations: (a) combined treatment of different waste streams, (b) ability to treat residues otherwise difficult to manage individually, (c) dilution of potentially toxic/inhibitory compounds, (d) resulting synergistic effects on hydrogen production, (e) optimization of the conditions for hydrogen production, (f) internal control of pH, and/or (g) optimization of the carbon/nitrogen ratio (Gioannis et al., 2013). Hydrogen and methane production and volatile solid removal efficiency were investigated in a two-stage mesophilic fermentation at various proportions of food waste and activated sludge (Liu et al., 2013). The hydrogen and methane yields reached 106.4 ml/g-VS and 353.5 ml/g-VS respectively at the food waste proportion of 85%, with the highest energy yield. The VS removal efficiencies of co-digestion were 10-77% higher than that of waste activated sludge fermentation. Only 0.1~3.2% of the COD in feedstock was converted into hydrogen, and 14.1~40.9% to methane, with the highest value of 40.9% in methane achieved at food waste proportion of 85%. Lin et al., (2013a) investigated the mesophilic anaerobic bio-hydrogen production from pulp & paper sludge (PPS) and food waste (FW), and the subsequent anaerobic digestion of the effluent for the methane production under thermophilic conditions by a two-stage process. The maximum hydrogen yield of 64.48 ml/g-VS_{added} and methane yield of 432.3 ml/g-VS_{added} were obtained when PPS and FW were applied with 1: 1 VS ratio as the feedstock. Removal efficiencies of SCOD for hydrogen and methane co-production reached 71%~87%. PH 4.8~6.4 and alkalinity 794~3316 mg $CaCO_3$/ L for H_2 fermentation, as well as pH 6.5~8.8 and alkalinity 4165~4679 mg $CaCO_3$/l for CH_4 fermentation, were achieved without any adjustment. This work indicated that anaerobic co-digestion of PPS and FW for hydrogen methane co-production was a stable, reliable and effective way for energy recovery and bio-solid waste stabilization by the two-stage mesophilic-thermophilic process. Their further studies showed that $NaOH/H_2SO_4$ pretreatment was suitable to enhance the hydrogen production in this two-stage mesophilic-thermophilic process (Lin et al., 2013b).

Thermophilic conditions also favor fermentative hydrogen production from these wastes. Previous studies showed that food waste produced around 81 ml/g-VS under thermophilic conditions, against 63 ml/g-VS observed under mesophilic conditions (Kim et al., 2008a). In spite of improved process performance under thermophilic conditions, it should be noted that they are energy consuming. If the energy for heating the fermentation system could be recovered from combined heat and power (CHP) system, thermophilic continuous processes could then be considered as sustainable.

Livestock Waste

The main types of animal manure include solid manure or farm yard manure; urinary waste i.e., slurry or liquid manure from livestock or poultry; and wastewater which is a collection of process water in farms, silage juices, feedlot runoff, bedding, disinfectants and liquid manure (Burton et al., 2003). Annual yield of animal manure is estimated to be more than 1500 million tons, including 1284 million tons of cattle manure and 295 million tons of pig manure across the 27 member states of the European Union (Holm-Nielsen et al., 2009). Manure will produce a major risk of air and water pollution, when manure is not well managed or treated. Nutrient leaching and pathogen contamination can lead to direct surface water damage. Moreover, manure can release up to 18% CO_2 equivalent and 37% CH_4, contributing to the greenhouse effect (Holm-Nielsen et al., 2009). Currently, agricultural biogas plants have been extensively used to co-digest manure and other waste biomass suitable for methane production. These large scale farm installations provide the necessary equipment to readily implement two-stage bioprocesses for co-production of hydrogen and methane in the future.

Anaerobic co-digestion of cow manure (CM) and waste milk (WM), produced by sick cows during treatment with antibiotics, was evaluated in a two-stage process under thermophilic condition (55°C) to determine the effect of WM addition on hydrogen and methane yields, VS removal, and energy recovery (Lateef et al., 2014). The highest specific H_2 and CH_4 yields and VS removal were 38.2 ml/g-VS, 627.6 ml/g-VS and 78.4%, respectively, in CM:WM 30:70. The results suggest that CM:WM 30:70 was optimum, however, due to limited amount of WM usually generated and long lag phase at this ratio which may make the process uneconomical, CM:WM 70:30 is recommended in practice. The hydrogen yield from dark fermentation of cow manure is obviously lower than that observed from crop residues (Guo et al., 2014; Cheng et al., 2012; Cheng and Liu, 2012a).

Low hydrogen yields of 14~29 ml/g-VS were also observed in other studies on dark fermentations of manure wastes (Xing et al., 2010; Yokoyama et al., 2007). It was concluded that the low hydrogen yield from pig slurry was mainly due to ammonium inhibition (Bonmati et al., 2001; Kotsopoulos et al., 2009). The ammonium concentration increased 3-fold over the initial value because of the decomposition of organic matter (Bonmati et al., 2001), while different livestock manures have been reported to contain 1.5~4 g N/l. Shock loading of manure wastes can produce severe inhibition of the whole biological anaerobic and hydrogen fermentation processes due to the high nitrogen content (Hobson et al., 1974; Salerno et al., 2006). The use of the co-digestion of manure wastes and carbohydrate-rich feed (e.g., crop residues) may provide a promising avenue for a much more stable and efficient fermentative hydrogen production.

Algae Biomass

Algae based fuels indicate great promise, directly related to the potential to produce more biomass per unit area in a year than any other form of biomass. Microalgae are a promising feedstock for biofuel and bioenergy production due to their high photosynthetic efficiencies, high growth rates and no need for external organic carbon supply (Lakaniemi et al., 2011).

Utilization of *Chlorella vulgaris* (a fresh water microalga) and *Dunaliella tertiolecta* (a marine microalga) biomass was tested as a feedstock for anaerobic H_2 and CH_4 production. Anaerobic serum bottle assays were conducted at 37°C with enrichment cultures derived from municipal anaerobic digester sludge. Low levels of H_2 were produced by anaerobic enrichment cultures, but H_2 was subsequently consumed even in the presence of 2-bromoethanesulfonic acid, an inhibitor of methanogens. H_2 production by the satellite bacteria was comparable from *D. tertiolecta* (12.6 ml/g-VS) and from *C. vulgaris* (10.8 ml/g-VS), whereas CH_4 production was significantly higher from *C. vulgaris* (286 ml/g-VS) than from *D. tertiolecta* (24 ml/g-VS). The high salinity of the *D. tertiolecta* slurry, prohibitive to methanogens, was the probable reason for lower CH_4 production (Lakaniemi et al., 2011).

The low hydrogen and methane yields in dark fermentation of raw microalgae biomass may be partly due to the recalcitrant cell walls of microalgae. Therefore, different strategies, which include chemical/physical pretreatments as well as bioaugmentation with cellulose-degrading microorganisms, have been used to enhance hydrogen production from microalgae biomass. A hydrogen yield of 28 ml/g dry weight from the microalgae *Laminaria japonica* was achieved by ball milling and heat treatment at 120°C for 30 min (Park et al., 2009), while Yang et al., (2011) observed a hydrogen yield of 27.3 ml/g-VS from lipid-extracted *Scenedesmus* biomass pretreated at 95°C for 30 min.

Considering that cellulose contributes to the cell wall recalcitrance of the microalgae *Chlorella vulgaris*, bioaugmentation with a cellulolytic and hydrogenogenic bacterium, *Clostridium thermocellum*, at different inoculum ratios, was investigated as a possible method to improve CH_4 and H_2 production of microalgae (Lü et al., 2013). A two-step process of addition of *C. thermocellum* first and methanogenic sludge subsequently could recover both hydrogen and methane. The hydrogen and methane yield from the two-step treatment was 53 and 321 ml/g-VS, respectively, which is significantly higher than the corresponding yield obtained from one-step treatment of simultaneous addition of *C. thermocellum* and methanogenic sludge (Lü et al., 2013). High hydrogen yields of 82 and 114 ml/g-VS were obtained from *C. vulgaris* using microalgae-associated bacteria and a thermophilic consortium at 60°C, respectively (Carver et al., 2011).

Effect of Process Conditions

In direct fermentation of biomass to hydrogen and methane, hydrogen production is often limited by the hydrolysis by cellulose-degrading microorganisms (Cheng and Liu, 2012a,b). In addition, optimal conditions for cellulose hydrolysis and hydrogen fermentation are different. For example, efficient cellulose hydrolysis has been reported near neutral pH, while the highest hydrogen yields from sugars are often observed at lower pH values ranging from 5.0 to 5.5 (Gioannis et al., 2013; Guo et al., 2010a). In this section, the effects of process conditions on two-stage system, mainly the effects on key hydrogen fermentation of biomass, are discussed in detail.

Inoculums for Hydrogen Production

In a two-stage system for co-production of hydrogen and methane, selection of inoculums in hydrogen fermentation stage attracted more attention while anaerobic sludge has been used in most of cases for fermentative methane production. Hydrogen production is significantly influenced by the inoculums type, as the fermentation end products are influenced by the bacterial metabolism. The inoculums used for fermentative hydrogen production include pure cultures of known species of hydrogen-producing bacteria and natural mixed cultures with or without pretreatment such as sludge, manure, compost piles and so on.

(1) Pure Cultures

It is easier for a system with pure cultures to detect metabolic shifts due to the reduced diversity of the biomass. Moreover, important information regarding operational conditions that promote high hydrogen yield and production rate can be identified in studies employing pure cultures. In general, various pure cultures have been usually investigated to produce hydrogen from a wide range of mono-substrates.

In the past decades, fermentative hydrogen production has been intensively investigated using pure cultures (Kotay and Das et al., 2007; O-Thong et al., 2008b, 2008c; Fabiano et al., 2002). A wide range of hydrogen-producing species have been isolated from natural mixed cultures, and they are more specifically mesophilic strains such as *Clostridium* (e.g., *C. pasteurianum, C. butyricum, C. saccharobutylicum*), *Enterobacter* (*E. aerogenes*) and *Bacillus*; and thermophilic or extremophilic strains such as *Thermoanaerobacterium* (*Thermoanaerobacterium thermosaccharolyticum*), *C. thermocellum*, *Bacillus thermozeamaize, Caldicellulosiruptor* (*Caldicellulosiruptor saccharolyticus*) (Chang et al., 2008; Hawkes et al., 2002; Karakashev et al., 2009; Ivanova et al., 2009).

As to microbial performances, mesophilic hydrogen producers obtain hydrogen yields ranging from 1.6 to 2.4 mol H_2/mol hexose for *C. beijerinckii, Enterobactor aerogenes, Enterobacter Cloacae, Clostridium butyricum* (Elsharnouby et al., 2013; Taguchi et al., 1995; Tanisho et al., 1994, 1995; Yokoi et al., 1997). Higher conversion yields were observed at high temperature for thermophilic or extremophilic strains. Maximum hydrogen yield of 2.53 mol H_2/mol hexose was observed for *T. thermosaccharolyticum* at a temperature of 60 ^{o}C (Quemeneur et al., 2012). For some other thermophilic microorganisms, such as *Thermotoga elfii, C. saccharolyticus, C. thermocellum, C. thermolacticum, C. thermobutyricum, and C. thermosaccharolyticum*, maximum hydrogen yields reached 1.5~3.3 mol H_2/mol hexose (Cao et al., 2009; Cheng and Liu, 2011; de Vrije et al., 2010; Li et al., 2014; Li and Liu, 2014; Ngo et al., 2012; O-Thong et al., 2008b, 2008c; van Niel et al., 2002). A hydrogen yield of 3.8 mol H_2/mol hexose, which is very close to the theoretical maximum level of 4.0 mol H_2/mol hexose, was observed for *Caldicellulosiruptosaccharolyticus* at 70 ^{o}C (Ivanova et al., 2009). This result may be partly resulted from the fact that hydrolysis and hydrogen production is favored at thermophilic conditions.

(2) Mixed Cultures

A variety of data in the literature reported hydrogen fermentation of waste biomass using natural mixed cultures as inocula and the need for inoculums pretreatment are among the most debated points in this process. In fermentative hydrogen production from different waste biomass such as agricultural waste, food waste and OFMSW, researchers use mixed microbial cultures for practical reasons, since a mixed culture system would be relatively easier to control, cheaper to operate, and capable of digesting a variety of feedstock materials (Li and Fang, 2007; Valdez-Vazquez et al., 2005). Anaerobic sludge from pilot or full scale anaerobic digesters, either with or without specific pretreatment, has been used in numerous studies as a supplier of a natural mixed anaerobic consortium. Among the most notable exceptions, Guo et al. (2014a), Fan et al., (2009), Han and Shin (2002), Lee et al., (2008, 2010a) and Cappai et al., (2009) used, respectively, cow dung compost, lesser panda manure, rumen microorganisms from the stomach of cows, an enriched culture from FW compost, and waste activated sludge (WAS) with no specific pre-treatment.

Although the use of mixed cultures for fermentative hydrogen production is practically more viable, obvious limitations arise from the coexistence of H_2-producing and consuming bacteria in nature (Gioannis et al., 2013). It thus necessitates a pre-treatment step by various methods to harvest hydrogen producers, based on their larger chances to survive when mixed cultures are treated by harsh conditions because of the ability of some hydrogen-producing bacteria, such as *Clostridium*, to sporulate as a reaction to adverse environment. To this regard, the main pretreatment methods available are heat-shock treatment (HST), aeration, acid treatment, freezing and thawing, as well as addition of specific chemicals.

Heat-shock treatment (HST), which is based on the ability of some bacterial species (e.g., *Bacillus* and *Clostridium*) to sporulate as a reaction to adverse environmental conditions, is however by far the most common approach to harvest hydrogenogenic bacteria among all options (Gioannis et al., 2013). Typically, HST is conducted around 100 $^{\circ}$C for durations of 15–120 min in order to suppress non-spore-forming bacteria, leaving spores of acidogenic bacteria that will germinate back to their active vegetative state when suitable growth conditions get re-established (Lay et al., 2003; Lin and Lay, 2004; Argun et al., 2008). In recent studies on food wastes and agricultural wastes, the pre-treatment conditions used ranged from 20 min at 80°C (Lee and Chung, 2010) to 30 min~2 h at 100°C (Lay et al., 2003; Guo et al., 2014).

Concerns about the energy balance of hydrogen fermentation using HST, however, make this biomass selection method controversial. Enrichment of hydrogen producers from inocula by HST is an energy-intensive practice. It should also be noted that continuous application of the inhibition method is typically required because of possible proliferation of new non-inhibited methanogens from non-sterile substrates (Gioannis et al., 2013). As a result, energy requirement will be further increased for fermentation of these wastes such as OFMSW, when HST is adopted to inhibit hydrogen-consuming microorganisms existing in these substrates. The degree of energy consumption can only be partially reduced by adopting temperatures as low as 75–85°C. All this presents a substantial hurdle for the development of cost-effective process.

With the aim of reducing costs and simplifying the process, researchers have tried to develop a process for hydrogen production from organic wastes using inoculum without pre-treatment (Chu et al., 2008; Lee et al., 2008, 2010a; Li et al., 2008b; Pan et al., 2008; Kim et al., 2011b). A method defined as biokinetic control, which suppresses the growth of hydrogen-consuming bacteria (e.g., methanogens) by maintaining environmental conditions such as low pH, appropriate temperatures or short HRTs causing the wash-out of methanogens (Valdez-Vazquez et al., 2005; Kim et al., 2011b).

In some studies, successful fermentative hydrogen production has been carried out with no inoculums added to the feed material (Gómez et al., 2006). In such cases, the evolution of the fermentation process relies on the indigenous microorganisms present in the waste substrate only. A stable performance (52.5–71.3 ml H_2/g-$VS_{removed}$) of hydrogen production was found in hydrogen-producing stage of a two-stage fermentation process for co-production of hydrogen and methane from an unsterilized mixture of OFMSW and slaughterhouse waste (Gómez et al., 2006)). Wang and Zhao (2009) developed a two-stage process, which was performed in a semi-continuous rotating drum reactor in which the indigenous mixed microbial cultures contained in food waste were used for fermentative hydrogen production. A maximum hydrogen yield of 65 ml H_2/g-VS was obtained in the hydrogen production stage, which was operated at an organic loading rate (OLR) of 22.65 kg-VS/m^3 d and a solids retention time (SRT) of 160 h. Another study also demonstrated a stable two-stage process combined hydrogen and methane production from household solid waste (Fountoulakis and Manios, 2013). The short HRT of 2 days applied in the first stage was enough to separate acidogenesis from methanogenesis and no additional control for preventing methanogenesis was necessary. Temperature control in the range 35~60 ^{o}C was selected as a biokinetic control strategy to optimize hydrogen production from food waste (Kim et al., 2011b). The optimal operational condition for both the hydrogen production yield and rate was found at an operating temperature of 50°C, with values of1.8 mol H_2/mol hexose and 369 ml/l h, respectively.

In addition, there are also studies in the literature where chemical inhibitors such as sodium-2-bromoethanesulfonate, 2-bromoethanesulphonic acid, methyl chloride, ethane, ethylene, iodopropaneacetylene, methyl fluoride, and chloroform, were added to inhibit hydrogen-consuming methanogens. Chemical inhibitors may be either specific or non-specific towards methanogens, which can include both H_2 consumers and other types of methanogens (Gioannis et al., 2013). The Coenzyme M (CoM) is involved in the terminal stage of methane biosynthesis, where the methyl group carried by CoM is reduced to methane by the methylCoM reductase. BES (sodium-2-bromoethanesulfonate), BESA (2-bromoethanesulphonic acid) and lumazine ($C_6H_4N_4O_2$), which are structural analogues of CoM specifically found in methanogens only but not in other bacteria or Archea, can competitively inhibit the methyl transfer reaction at the terminal reducing stage of methane formation from hydrogen and CO_2. Acetylene was used as a non-specific inhibitor of methanogens, while ethylene is recommended as a reversible selective inhibitor of methanogenesis and methanogenic activity has been reported to completely recover after ethylene removal. Chloroform ($CHCl_3$) is known to inhibit the function of corrinoid enzymes and the methylCoM reductase. $CHCl_3$ can inhibit both acetoclastic and hydrogenotrophic methanogens. However, $CHCl_3$ investigation has been found not only to inhibit the activity of methanogenic Archaea but also that of homoacetogenic bacteria and acetate-consuming sulfate-reducing bacteria.

Successful hydrogen fermentation process has been carried out using different pretreatment methods of inoculums. Further studies are needed to the development of the strategy for H_2-consumers inhibition based on capital and operational costs, feasibility and complexity of the process layout; time needed for inoculum stabilization, effectiveness over the entire fermentation process, and degree of compatibility with further process steps such as methane production from hydrogen fermentation residues.

pH

The effect of pH on fermentative hydrogen production is very controversial in the literature, while it is concluded that neutral pH is favorable for methane production by anaerobic digestion. The optimal pH for hydrogen production from carbohydrates is in the range of 5.2 to 7.0 (Li and Fang, 2007). Optimal hydrogen production is found to take place with a pH of 5.0~6.0 for food wastes, whereas a neutral pH is preferred for crop residues and animal manure wastes (Guo et al., 2010a; Kim et al., 2004; Kim et al., 2006a; Li and Chen, 2007; Zhang et al., 2007). Low hydrogen yields have been reported at initial pH values 5 and 9, and initial pH below 5 has often inhibited hydrogen production (Li and Chen, 2007; Pan et al., 2010). The optimal initial pH is determined by the H_2 producing bacterial community. However, most studies on pH effects have been finished under conditions without pH control. Optimal initial pH for hydrogen production from biomass has been between 6 and 7 with enrichment cultures from cow dung compost (Fan et al., 2006; Zhang et al., 2007), 5.5 with *Clostridium butyricum* (Pattra et al., 2008), 7.0~7.5 with *C. butyricum* AS1.209 (Li and Chen, 2007) and 8.0 with dairy manure bacteria (Pan et al., 2010). These studies give only an indication of suitable initial pH, but not the optimal hydrogen production condition.

On-line pH control used in other studies may further elucidate the effect of pH (Nissila et al., 2014; Elsharnouby et al., 2013). The effect of pH on the conversion of glucose to hydrogen by a mixed culture of fermentation bacteria was evaluated at pH ranging from 4.0 to 7.0 and the optimal pH was found to be 5.5 (Fang and Liu, 2002). Fan et al., (2008) investigated the effect of the operational pH on bio-H_2 production. In their study, the bio-pretreated corn stalk of 15 g/l was fermented at different operating pH ranging from 4.5 to 7.0 in a 5-L CSTR and the optimal hydrogen production was also carried out at fixed pH of 5.5 (Fan et al., 2008). Some other studies revealed that optimal fixed pH for hydrogen production from various substrates has been between 5.0 and 6.0 with pure or mixed cultures (Lin and Chang, 1999; Kumar and Das, 2000).

Low pH affects the hydrogenase activity, which is regarded to as a important factor elucidating the effect of pH on hydrogen fermentation (Khanal et al., 2004; Nazlina et al., 2011). The metabolic pathways involving acetate and butyrate production appear to be favored at pH ranging from 4.5 to 6.0, while neutral or higher pHs are believed to promote ethanol and propionate production (Guo et al., 2010a; Gioannis et al., 2013). Solventogenesis is assumed as a detoxification method of the biomass to avoid inhibitory effects caused by high VFA contents and associated low pH in the broth (Gioannis et al., 2013). However, other researchers observed a shift to solventogenesis at pH levels above 5.7, due to the synthesis or activation of the enzymes required for solvent production (Khanal et al., 2004). Previous studies (Nazlina et al., 2011) indicated that decreasing pH below 6.0 increasingly promoted lactate formation, with an associated negative influence on the hydrogen yield. In addition, the pH may also affect the degree of biomass activity, with values less than 6 capable of significantly inhibiting sulfate-reducing and methane-producing microorganisms. Mantaining low pH and short HRTs is thus used as an effective approach of biokinetic control to build successful fermentative hydrogen production with untreated mixed cultures (such as activated sludge) as inoculums (Gioannis et al., 2013).

Temperature

Hydrogen fermentation of pure substrates has been widely studied with mesophilic (20 to 40°C), thermophilic (50 to 65°C) and hyperthermophilic (> 70°C) microorganisms. When compared with mesophilic fermentation of acid hydrolyzed wheat powder (Cakir et al., 2010) and heat- and enzyme pretreated bagasse (Chairattanamanokorn et al., 2009), increased hydrogen yields and rates and shortened lag time were carried out in their thermophilic fermentation individually. Temperature influenced the soluble metabolite distribution, and lactate production dominated at other temperatures than 28°C (Nissila et al., 2012). A number of studies reported higher hydrogen yields were obtained under thermophilic condition than mesophilic condition (Section "Hydrogen and methane potential of different biomass").

Hydrogen Partial Pressure

It is concluded that hydrogen partial pressure is a restrictive factor in the process of the fermentation of organic waste (Guo et al., 2010a). The oxidation of reduced components such as Long Chain Fatty Acids to VFAs, concomitantly with hydrogen production, is the result of a low biohydrogen level in the medium because reactions are thermodynamically unfavorable (Li, 2009). Additional hydrogen production could also derive from the degradation of acetate (Angenent et al., 2004). This conversion is thermodynamically unfavorable at moderate temperatures and the reaction is therefore quite sensitive to hydrogen concentration. Furthermore, the inverse reaction (i.e., homoacetogenesis) is rather favored in the fermentation process and partly reduces the performance of bioreactors through the accumulation of acetate in the medium. The increase of the hydrogen concentration in the medium may affect not only biohydrogen production, but also a shift of metabolic pathways towards solventogenesis has been observed, i.e., the accumulation of lactate, ethanol, acetone and butanol (Levin et al., 2004).

Many studies reported that reducing hydrogen pressure by using inert gas sparging increased hydrogen yield, although the degree of improvement is highly variable between studies, ranging from relatively little change to a significant doubling of hydrogen yield to 3.9 mol H_2/mol hexose (Kim et al., 2006b; Mandal et al., 2006). Chou et al., (2006) reported that the biohydrogen production increased from 1.8 ml/l reactor to 6.1mL/l reactor while the stirring speed was increased from 20 to 100 rpm in a 100 L pilot bioreactor treating brewery grains.

Several other alternatives exist to improve gas extraction, including gas sparging and biohydrogen stripping from reactor headspace by membrane absorption (Chou et al., 2006). The main disadvantage of these techniques is that the sparging gas dilutes the hydrogen content and produces a further decrease in separation efficiency.

Pretreatment of Waste Biomass

The cost, availability, carbohydrate content and biodegradability of the material are very important factors that affect the economic lly feasibility of bioconversion of biomass to hydrogen and methane (Kapdan and Kargi, 2006). Lignocellulosic waste is an ideal resource for renewable hydrogen and methane production due to its high annual production; however, it is quite difficult to produce hydrogen and methane from lignocellulosic waste due to its recalcitrant structure (Levin et al., 2009; Lu et al., 2009). Depending on the composition and accessibility of biomass, it may require pretreatment and/or hydrolysis prior to use for dark hydrogen/methane fermentation.

The biodegradability of lignocellulosic biomass is limited by not only crystallinity of cellulose, but also degree of polymerization (DP), moisture content, available surface area and lignin content (Chang and Holtzapple, 2000; Koullas et al., 1992; Laureano-Perez et al.,, 2005; Puri, 1984). Different pretreatments have an effect on one or more of these aspects, as showed in Table 2.

The table shows the importance of improving the surface area, which is one of the major approaches of a pretreatment by solubilization of the hemicellulose and/or lignin and/or altering the lignin (Hendriks and Zeeman, 2009; Zeng et al., 2007). The different effects of several pretreatments on lignocelluloses to improve its digestibility are reviewed in detail by Hendiriks and Zeeman (2009).

It can be concluded that concentrated acids, wet oxidation and solvents pretreatments are effective, but too expensive compared to the value of sugar (Fan et al., 1987; Mosier et al., 2005b). Liquid hot water pretreatments, lime pretreatment, ammonia based pretreatments and steam pretreatment are concluded to be methods with high potentials. The main effects are dissolving hemicellulose and alteration of lignin structure, providing an improved accessibility of the cellulose for hydrolytic enzymes and/or microorganisms.

Table 2. Effects of the different pretreatments on the physical/chemical composition or structure of lignocellulose (Hendriks and Zeeman, 2009; Mosier et al., 2005b)

	Increase accessible surface area	Decrystallization cellulose	Solubilization hemi-cellulose	Solubilization lignin	Formation furfural/HMF	Alteration lignin structure
Mechanical	+	+				
ST/SE	+		+	-	+	+
LHW (batch)	+	ND	+	-	-	-
LHW (flow through)	+	ND	+	+/-	-	-
Acid	+		+	-	+	+
Alkaline	+		-	+/-	-	+
Oxidative	+	ND		+/-	-	+
Thermal + acid	+	ND	+	+/-	+	+
Thermal + alkaline (lime)	+	ND	-	+/-	-	+
Thermal + oxidative	+	ND	-	+/-	-	+
Thermal + alkaline + oxidative	+	ND	-	+/-	-	+
Ammonia (AFEX)	+	+	-	+	-	+
CO_2 explosion	+		+			

Notes: +: major effect. -: minor effect. ND: not determined. LHW: Liquid hot water. AFEX: Ammonia fiber explosion.

An important aspect is the choice of the bio-fuel to produce. The differences between the ethanol and methane (or hydrogen & methane) conversion efficiencies are not large anymore, because nowadays also yeast strains capable of converting C5 and C6 sugars to ethanol are available (Hendriks and Zeeman, 2009, Kuyper etal., 2005). An advantage of the biogas production, compared to ethanol production, is the high tolerance for inhibiting compounds like furfural and HMF by the mixed cultures.

The effect of the pretreatments is however very dependent on the biomass composition and operating conditions. All these pretreatments have their advantages and shortcomings and future study is needed for optimization.

It should be mentioned that inhibitory compounds, that include furfural, HMF and carboxylic acids, are likely produced in acid, alkaline and steam explosion pretreatments. Furfural and HMF are oxidation products of glucose and xylose, respectively, while other phenolic compounds result from the partial-degradation of lignin (Cao et al., 2010; Nissila et al., 2014). A number of studies showed that dark fermentative hydrogen production may be inhibited by these compounds (Cao et al., 2010; Jung et al., 2011). Inhibitory compounds have been decreased before dark fermentation with, e.g., charcoal, activated carbon, cation exchange resin, overliming (Jonsson et al., 2013; Lee et al., 1999; Sanio et al., 2011). Previous studies showed that no hydrogen was observed directly from the acid hydrolyzate of rice straw, and the hydrogen yield was increased by detoxification with $Ca(OH)_2$ (overliming) to remove furfural and parts of VFAs (Chang et al., 2011).

Bioreactor Configuration and Operation

Different reactor configurations have been used to treat waste biomass for hydrogen production, mostly consisting of lab-scale (100–500 ml) bottles and stirred reactors of 2–10 l, operated under batch, semi-continuous or continuous conditions (Gioannis et al., 2013; Guo et al., 2010a). Because of the advantage of being easily operated and flexible, lab-scale batch reactors were widely used for determining the biohydrogen potential of organic substrates (Fan et al., 2006; Pakarinen et al., 2008). However, in an industrial context, for practical reasons of waste management and for economic considerations, continuous bioprocesses are recommended. To date, although no full scale reactor for hydrogen fermentation has been set up, it is expected that bioreactor design and system configuration will be similar to biogas plants: only the operational parameters may vary between these two anaerobic applications.

Based on the extensive experience acquired in biogas plants treating agricultural wastes, especially in Germany, the most probable bioreactor for biohydrogen production would be a continuous stirred tank reactor (CSTR)(Weiland et al., 2006). The anaerobic bottles with or without manual/mechanical shaking and CSTR are the most common systems used in lab-scale hydrogen fermentation research on substrates such as swine manure and food waste (Kotsopoulos et al., 2009; Shin et al., 2006; Zhu et al., 2009). Successful application of anaerobic sequencing batch reactor (ASBR), rather than CSTR, for food waste conversion has also been reported (Kim et al., 2008b). The maximum hydrogen yield of 1.12 mol H_2/mol hexose was carried out at a SRT of 126 h and a HRT of 33 h. Alzate-Gaviria et al., (2007) obtained a yield of 99 N ml/g-VS removed in a packed bed reactor (PBR). In addition, a pilot-scale plug-flow reactor was designed to investigate kitchen waste in hydrogen conversion (Jayalakshmi et al., 2009). This reactor is cylindrical in shape and kept at a $20°$ angle to the horizontal to facilitate movement of the substrate. A screw inside the reactor is used to push the substrate from the inlet at the bottom to the outlet at the top. Batch mode is preferred for a successful start-up (Jayalakshmi et al., 2009).

The OLR and HRT/SRT is considered as important operational parameters in fermentative hydrogen production, which affect a number of issues including VFA accumulation, pH changes, substrate degradation, the type of active microbial population as well as their metabolic pathways. Since no biomass recycle was used for hydrogen-producing reactors in most of previous studies, HRT and SRT coincide. In most continuous or semi-continuous operations of stirred reactors, HRT values were between 21 h and 4 d (Lee and Chung, 2010; Lee et al., 2010), while the reported OLRs values varied from 8 to 38 kg-VS/m^3 d (Hong and Haiyun, 2010; Chu et al., 2008) or from 20 to 64 kg-COD/m^{3}d (Li et al., 2008b; Chu et al., 2008).

The influence of HRT and OLR on hydrogen yield is controversial in the literature. Hydrogen yield was more than doubled when the HRT of a semi-continuous system was prolonged from 2 to 5 d and the OLR was reduced from 10 to 8 kg-VS/m^3 d (Shin and Youn (2004)). The decreased OLR was observed to prevent VFAs to accumulate in excess of 20,000 mg-COD/l. A significant reduction in VS removal and hydrogen yield was also found as OLR progressively adjusted from 15.10 to 37.75 kg-VS/m^3 d and SRT decreased from 10 to 6.6 d due to reduced time span of substrate hydrolysis (Wang and Zhao, 2009). Furthermore, increased OLRs were found to result in a metabolic shift from acetate and butyrate production to synthesis of propionate and lactate.

In another study, a marked increase in the hydrogen yield was observed when OLR was decreased from 30.20 to 22.65 kg-VS/m^3 d (corresponding to an increase in SRT from 5 to 6.6 d), while a further decrease in OLR from 22.65 to 15.10 kg-VS/m^3 d (corresponding to an increase in SRT from 6.6 to 10 d) only slightly improved hydrogen production. Hong and Haiyun (2010) reported that long HRTs (8.9 d) were found to improve hydrogen yield in semi-continuous digestion of FW and dewatered WAS. A prolonged SRT of 21 days (corresponding to an OLR of 11 g-VS/kg d) were also adopted by Valdez-Vazquez et al., (2005) and the hydrogen production yield reached 165 and 360 N ml/g-VSrem under mesophilic and thermophilic conditions, respectively.

In order to improve the substrate degradation and energy recovery, a two-stage systems for co-production of hydrogen and methane is recommended for treating substrates such as agricultural stalk waste, livestock waste and food waste (Cheng and Liu, 2012a; Cheng et al, 2012; Wang and Zhao, 2009; Koutrouli et al., 2009). Fast-growing acidogens are dominant in the first stage of hydrogen fermentation in this system, while slow-growing acetogens and methanogens are the main microorganisms in the second stage for bioconversion of VFAs into methane. Previous study reported that only 5.78% of the influent COD was fermented into hydrogen in the first stage, against 82.18% of COD converted to methane in the second stage (Wang and Zhao, 2009). Successful association of reactors for hydrogen and methane production from food waste was reported (Chu et al., 2008). In this study, thermophilic hydrogen-producing reactor was operated at a short HRT (31 h) and acidic pH (5.5) to prevent methanogenic activity in the acidogenic stage, while methane fermentation was then carried out at 35 °C, neutral pH, 120 h HRT. High biogas yield of 464 ml/g-VS with methane content of 70~80% was observed thanks to the hydrolytic activity in the first step and a HRT of 5 days was enough for the methane stage, against a HRT of 10~15 days in traditional anaerobic digestion (Li et al., 1999). In a recent study, a remarkable three-stage system coupling two-stage batch hydrogen fermentation with continuous methane production has been developed for treating stalk wastes (Cheng et al., 2012). In the first hydrogen-producing phase, corn stalk was efficiently degraded by thermophilic bacteria and overall utilization ratio of substrate reached 70%. The liquid effluents of hydrogen fermentation were then digested in a methane-producing UASB reactor operated at 35 °C and a short HRT of 10 h.

Energy Recovery of Two-Stage System

Kim et al., (2013) evaluated bioenergy recovery in a two-stage batch thermophilic fermentation using sewage sludge (SS) and untreated rice straw (RS) as raw materials for co-production of hydrogen and methane. Overall, volatile solids removal (60.4%) and total bioenergy yield (8.8 kJ/g-VS) for the two-stage system were 37.9% and 59.6% higher, respectively, than the one-stage methane fermentation system. The total energy recovery by co-production of hydrogen and methane from cornstalk based on direct microbial conversion (DMC) with thermophilic baterica with a strong cellulose-degrading ability, e.g., *C. thermocellum* reached 70.0%, corresponding to a bioenergy yield of 8.3 kJ/g-cornstalk (Cheng and Liu, 2012a). Hydrogen and methane production, energy yield, and volatile solid removal efficiency were investigated in two-stage mesophilic fermentation at various proportions of food waste and activated sludge (Liu et al., 2013). The highest energy yield reached 14.0 kJ/g-VS at the food waste proportion of 85%. The energy yields from two-stage dark fermentation of OFMSW/food waste are relatively higher than that observed from crop residues, which may be because of recalcitrant structure of the latter. Pretreatment of substrates is proved as useful method to improve energy recovery of two-stage fermentation. A two-stage process for extreme thermophilic hydrogen and thermophilic methane production from wheat straw hydrolysate was investigated in up-flow anaerobic sludge bed (UASB) reactors (Kongjan et al., 2011). The energy conversion efficiency was dramatically increased from only 7.5% in the hydrogen stage to 87.5% of the potential energy from hydrolysate, corresponding to total energy of 13.4 kJ/g-VS. A two-step process of addition of *C. thermocellum* first and methanogenic sludge subsequently could recover both hydrogen and methane from algae biomass. The overall energy yield reached 13.4 kJ/g-VS, which is 9.4% higher than the corresponding yield obtained from one-step treatment of simultaneous addition of *C. thermocellum* and methanogenic sludge (Lü et al., 2013). Anaerobic co-digestion of cow manure (CM) and waste milk (WM) was evaluated in a two-stage process under thermophilic condition (55 °C) (Lateef et al., 2014). The highest energy yield reached up to 25.5 kJ/g-VS in CM:WM 30:70.

The production costs and electricity production for hydrogen-only fermentation, methane-only fermentation, and hydrogen/methane fermentation have been evaluated by Lee and Chung (2010). According to the economic evaluation, little difference in production costs was observed between the three systems. Had the hydrogen yield been 1.0 mol H_2/mol glucose, however, two-phase hydrogen/methane fermentation would have been expected to produce 7~9 times more electricity than hydrogen-only fermentation and approximately 10~12% more electricity than methane-only fermentation. When a hydrogen yield of 1.82 H_2 mol H_2/mol hexose is applied, energy production can increase 25% with two-phase hydrogen/methane fermentation compared with methane-only fermentation.

Pilot Scale Studies of Two-Stage System

To our best knowledge, there is no data on full-scale hydrogen fermentation or hydrogen- and methane two-stage fermentation plants until now, while anaerobic digestion for methane production is a mature technology. Some experience has recently been obtained on pilot-scale reactors.

Optimization of a two-phase thermophilic anaerobic process treating biowaste for hydrogen and methane production was carried out at pilot scale using two stirred reactors (CSTRs: HR, 200 L; MR, 760 L) and without any physical/chemical pre-treatment of inoculums (Cavinato et al., 2011). During the experiment the hydrogen production at low hydraulic retention time (3d) was tested, both with and without reject water recirculation and at two organic loading rates (16 and 21 kg TVS/m^3d). The optimal yield was obtained with recirculation where the pH reached an optimal value (5.5) thanks to the buffering capacity of the recycle stream. The specific gas production of the first reactor was 51 l/kg-VS_{fed} and H_2 content in biogas 37%. The mixture of gas obtained from the two reactors met the standards for the biohythane mix only when lower loading rate were applied to the first reactor, with a composition of 6.7% H_2, 40.1% CO_2 and 52.3% CH_4 the overall SGP being 0.78 m^3/kg-VS_{fed}.

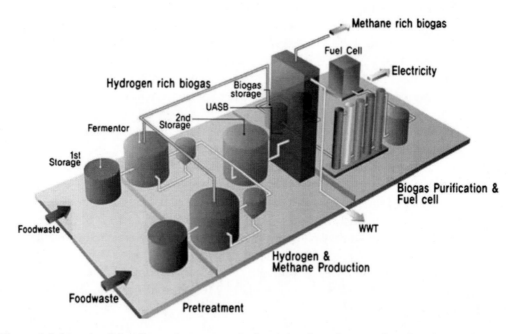

Figure. 1 Schematic of the pilot-scale two-stage hydrogen/methane fermentation plant (Lee and Chung, 2010).

A pilot-scale anaerobic SBR with 150 l working volume was used to treat food waste by Kim et al., (2010). Schematic of the pilot-scale two-stage hydrogen/methane fermentation is shown in Figure 1. A hydrogen yield of 0.5 mol H_2/mol hexose was observed when the reactor was operated at 35 °C and an HRT of 36 h. In another study on co-production of hydrogen and methane from food waste in a pilot-scale two-phase system, a hydrogen yield of 1.82 mol H_2/mol hexose was carried out in a 500 l CSTR operated at pH 5.5, 33 °C and an optimal HRT of 21 h (Lee and Chung (2010)). Subsequently, hydrogen fermentation residues were converted into methane in a 2000 l UASB operated at pH 7.4, 37 °C, and an optimal HRT of 3.9 d. The methane fermentation UASB allowed for a production rate of ~7 l/l-d with a methane content of 81~83%.

Two-stage hydrogen/methane fermentation has significantly greater potential for recovering energy than conventional hydrogen-only fermentation and methane-only fermentation. Many challenges must be addressed, however, such as the establishment of technical reliability, economical feasibility and the verification of a full-scale plant.

COMPARISON WITH OTHER HYBRID PROCESSES

Dark-Fermentation

As mentioned above, the residual organic content in the effluent from the hydrolytic and hydrogen-producing stage, which is mainly in the form of the soluble products, may be efficiently converted into methane in a second digester. The advantages of co-production of hydrogen and methane by two-stage dark fermentation are very clear, although some shortcomings, such as low hydrogen yield, are observed in this process (Table 3). When compared to conventional methanogenesis, an increase of 12-25% in energy production can be carried out in two-stage fermentation for co-production of hydrogen and methane. The energy recovery of this two-stage process is much higher than that achieved in individual hydrogen fermentation (Cheng et al., 2012; Cheng and Liu, 2012a). Standard anaerobic digestion techniques and engineering experience used to convert organic acid to methane are available, and the nature of light-independent process really simplifies the design of bioreactor and operation. All this provides important basis for future application of this process. Alternatively, the organic content remained in the effluent from the first stage may be further exploited through other biological processes such as photo-fermentation and microbial electrolysis cells (MEC) as follows.

Photo-Fermentation

Further hydrogen production may also be carried out by combining dark fermentation with photo-fermentation. Purple non-sulfur photosynthetic bacteria are able to use short-chain organic acids as electron donors to generate hydrogen through a light-driven reaction (Gest and Kamen, 1949; Hillmer and Gest, 1977). If cost effective photobioreactors were available, the two-stage dark- and photo fermentation process using mixed microflora would be a potential avenue as it has a theoretical maximal molar H_2 yield of 12 mol H_2/mol hexose converted in the two-stage process, although conversion efficiencies are substrate dependent (Tao et al., 2009). Fang et al., (2005) showed for the first time that mixed photofermentative microflora enriched from a reservoir could be used to ferment acetate and butyrate effectively into hydrogen. A two-stage batch process including dark- and photo-fermentation was developed to produce hydrogen from food waste (Zong et al., 2009). Organic acids in the effluent of hydrogen-producing stage were further consumed by *R. Sphaeroides* ZX-5 during 168-h photo-fermentation. The hydrogen yields achieved in these two-stages were 1.77 and 3.63 mol/mol hexose, respectively, with an overall yield of 5.4 mol H_2/mol hexose, equivalent to a conversion efficiency of 45%. (Zong et al., 2009).

A number of other studies showed that the organic acids, principally acetate and butyrate, produced during a first stage dark hydrogen fermentation of sugars, were successfully converted to hydrogen in a second photo fermentation, thus increasing the overall hydrogen yield, although yields were well below stoichiometric level (Nath et al., 2008; Chen et al., 2008).

Table 3.Advantages and disadvantages of different hybrid processes

First stage (for Bio-H$_2$)	Second stage (for Bio-CH$_4$)	Advantages	Disadvantages
Dark fermentation	Dark fermentation	➢ 12-25% increase in energy production compared to conventional methanogenesis ➢ Light-independent process ➢ Standard anaerobic digestion techniques used to convert organic acid to methane	➢ Low hydrogen yield ➢ Low production rate ➢ large reactor volume ➢ Two different fuels produced
	Photofermentation	➢ Has the ability to fix N$_2$ from atmosphere ➢ A wide spectral light energy can be used by these bacteria ➢ Can use different organic wastes	➢ inhibitory effect of O$_2$on nitrogenase ➢ Low light conversion eficiency ➢ Costly H$_2$-imperable photobioreactor ➢ Scaling up problems
	Microbial electrolysis cells	➢ Ehanced hydrogen recovery (theory complete substrate conversion to H$_2$, 12 mol H$_2$/mol hexose) ➢ Light-independent process	➢ External energy supply ➢ Increased system complexity ➢ Added energy required ➢ Low current densities, large surface area electrodes required

As shown in Table 3, the most severe disadvantage is the fact that light conversion efficiencies, with varied light sources including solar radiation and tungsten lamps, are quite low (only 1-5%) (Dasa and Veziroglub, 2008; Gioannis et al., 2013; Hallenbeck, 2009). The majority of captured light is wasted as heat, even at moderate light intensities (Dasa and Veziroglub, 2008). This, combined with the potentially high cost of photo-bioreactors and limitations in light penetration due to turbidity, presents very significant hurdles for the development of an economically viable process based on photo-fermentation.

Microbial Electrolysis Cells

Bio-electrochemically assisted microbial reactors have also been proposed as methods to couple with hydrogen fermentation. Among these, hydrogen production in microbial electrolysis cells (MECs), which was initially developed by Liu et al., (2005), is a process in which that electrical energy is added to drive conversion of organic acids to hydrogen. In this system, the electrons produced by the degradation process are transferred through the external circuit to a cathodic compartment. Protons are transferred through the ion exchange membrane, which separates the two compartments. Hydrogen production is then carried out in a process that electrons reduce the protons generated by the degradation process in the cathodic compartment of MEC (Liu et al., 2005; Logan et al., 2008; Jeremiasse et al., 2010). For electrochemically-driven hydrogen production in MECs, the minimum theoretical applied voltage is 100 mV. In practice, the minimum applied voltage to produce hydrogen from the bio-electrolysis of acetate has been found to be more than 200 mV because of electrode over potential (Liu et al., 2005; Hallenbeck, 2009).

Although the information on combination of dark fermentation and MECs for hydrogen production from food waste (FW) is relatively rare, the MEC has been shown to be able to generate hydrogen from a variety of complex substrates besides pure volatile acids (VFAs) (Guo et al., 2010b; Manuel et al., 2010; Cheng and Logan, 2011), including lignocellulosic materials (Lalaurette et al., 2009), cellulose (Wang et al., 2011), and wheat powder (Tuna et al., 2009). When considering that the nature of the metabolic products of acidogenesis is similar for various kinds of organic wastes, the results of the above studies present a promising alternative for enhanced hydrogen production from all waste biomass by combining with dark hydrogen fermentation and MEC process. The main disadvantages of this process are external energy supply, increased system complexity and large surface area electrodes required, which bring substantial barriers for economical feasibility of this process.

CONCLUSION

This chapter reports recent findings on co-production of hydrogen and methane from waste biomass by two-stage dark fermentation. Different kinds of waste biomass have been considered in the present review, which includes crop residues, animal manure, OFMSW and food waste etc. It is shown that all biomass possess great potential as a substrate for hydrogen and methane production by dark fermentation. Relatively higher hydrogen and methane yield as well as energy recovery were observed in fermentation of food waste and OFMSW than that obtained from crop residues. The two-stage processes are not only influenced by the composition of the organic wastes, but also they are highly dependent of the operational parameters. Key operating conditions such as low pH, low partial pressure of hydrogen, high temperature and acclimated microbial communities are recommended to carry out successful fermentative hydrogen production. The association of a hydrogen fermentor with a methanogenic reactor is proved to achieve the efficient conversion of biodegradable organic matter to bioenergy, which is indicated by much higher energy yields than that observed in single hydrogen fermentation.

Although the potential of co-production of hydrogen and methane by two-stage dark fermentation and technical feasibility of treating simple substrates and complex waste biomass has been demonstrated by a number of studies, the technology of hydrogen fermentation in the first stage still appears to be in an early stage, especially for complex waste substrates. To our best knowledge, there is no full scale bio-hydrogen plant in operation now. To improve economic feasibility of the two-stage dark fermentation process for co-production of hydrogen and methane, the following points require immediate attention:

(a) When considering that the performances of the biological processes are related not only to the operating conditions, but also, to the composition of the waste substrates, future research is needed to better understand the effect of feedstock composition, to predict bioreactor performances and optimize the co-digestion system.

(b) An expanded feedstock range and sources and the use of mixed cultures or the indigenous biomass in the substrate as the inoculums would significantly improve the chance of successful-development of the process.

(c) Research attention is needed to explore the possibility of optimizing the process performance by appropriately adjusting process configuration and operation, with no need for external control of the operating variables or for application of severe conditions.

(d) In two-stage processes, the major bottleneck lies on the dark hydrogen fermentation processes, particularly for lignocellulosic biomass. Improvement of these processes, either exploring economically feasible pretreatments to increase accessibility of lignocellulosic substrates or developing bioaugmentation strategies such as co-cultures of efficient cellulose-degrading microorganisms (e.g., *C. thermocellum*) and high hydrogen producers, surely will improve overall biogas yield (including hydrogen and methane) as well as economy of the process.

(e) Attention should be directed toward improving overall performance and ultimate bioenergy recovery in the two-stage dark fermentation process from various types of substrates rather than only focusing on high hydrogen yield and production rate of the first hydrogen fermentation stage.

REFERENCES

Alzate-Gaviria, L.M., Sebastian, P.J., Perez-Hernandez, A., Eapen, D., (2007). Comparison of two anaerobic systems for hydrogen production from the organic fraction of municipal solid waste and synthetic wastewater. *Int. J. Hydrogen Energy* 32, 3141-3146.

Angelidaki, I., Ahring, B., (1994). Anaerobic thermophilic digestion of manure at different ammonia loads: effect of temperature. *Water Res.* 28(3), 727-731.

Angenent, L.T., Karim, K., Al-Dahhan, M.H., Wrenn, B.A., Domıguez-Espinosa, R., (2004). Production of bioenergy and biochemicals from industrial and agricultural wastewater. *Trends Biotechnol.* 22(9), 477-485.

Argun, H., Kargi, F., Kapdan, I.K., Oztekin, R., (2008). Batch dark fermentation of powdered wheat starch to hydrogen gas: effects of the initial substrate and biomass concentration. *Int. J. Hydrogen Energy* 33, 6109-6115.

Balat, M., Balat, M., Kırtay, E., Balat, H., (2009). Main routes for the thermo-conversion of biomass into fuels and chemicals. Part 1: pyrolysis systems. *Energy Convers. Manage.* 50:3147–3157.

Burton, C., Turner, C., (2003). Manure management & treatment strategies for sustainable agriculture. Maulden, Bedford, UK: Turn-Around Typesetting Ltd.Bonmati, A., Flotats, X., Mateu, L., Campos, E., (2001). Study of thermal hydrolysis as a pretreatment to mesophilic anaerobic digestion of pig slurry. *Water Sci. Technol.* 44(4), 109-116.

Cakir, A., Ozmichi, S., Kargi, F., (2010). Comparison of bio-hydrogen production from hydrolyzed wheat starch by mesophilic and thermophilic dark fermentation. *Int. J. Hydrogen Energy* 35(24), 13214-13218.

Cao, G.L., Ren, N.Q., Wang, A.J., Guo, W.Q., Xu, J.F., Liu, B.F., (2010). Effect of lignocellulose-derived inhibitors on growth and hydrogen production by

Thermoanaerobacterium thermosaccharolyticum W16. *Int. J. Hydrogen Energy* 35(24), 13475-13480.

Cao, G.L., Ren, N.Q., Wang, A.J., Lee, D., Guo, W.Q., Liu, B.F., (2009). Acid hydrolysis of corn stover for biohydrogen production using *Thermoanaerobacterium thermosaccharolyticum* W16. *Int. J. Hydrogen Energy* 34, 7182-7188.

Carver, S.M., Hulatt, C.J., Thomas, D.N., Tuovinen, O.H., (2011). Thermophilic, anaerobic co-digestion of microalgal biomass and cellulose for H_2 production. *Biodegradation* 22, 805-814.

Cavinato, C., Bolzonella, D., Fatone, F., Cecchi, F., Pavan, P., (2011). Optimization of two-phase thermophilic anaerobic digestion of biowaste for hydrogen and methane production through reject water recirculation. *Bioresour. Technol.* 102, 8605-8611.

Chairattanamanokorn, P., Penthamkeerat, P., Reungsang, A., Lo, Y.C., Lu, W.B., Chang, J.S., (2009). Production of biohydrogen from hydrolyzed bagasse with thermally preheated sludge. *Int. J. Hydrogen Energy* 34(18), 7612-7617.

Chandra, R., Takeuchi, H., Hasegawa, T., (2012). Methane production from lignocellulosic agricultural crop wastes: a review in context to second generation of biofuel production. *Renew. Sustain. Energy Rev.* 16(3), 1462-1476.

Chang, J.J., Wu, J.H., Wen, F.S., Hung, K.Y., Chen, Y.T., Hsiao, C.L., Lin, C.Y., Huang, C.C., (2008). Molecular monitoring of microbes in a continuous hydrogen-producing system with different hydraulic retention time. *Int. J. Hydrogen Energy* 33(5), 1579-1585.

Chang, A.C.C., Tu, Y.H., Huang, M.H., Lay, C.H., Lin, C.Y., (2011). Hydrogen production by the anaerobic fermentation from acid hydrolysed rice straw hydrolysate. *Int. J. Hydrogen Energy* 36(21), 14280-14288.

Chen, C.Y., Yang, M.H., Yeh, K.L., Liu, C.H., Chang, J.S., (2008). Biohydrogen production using sequential two-stage dark and photo fermentation processes. *Int. J. Hydrogen Energy* 33, 4755-4762.

Cheng, S., Logan, B.E., (2011). High hydrogen production rate of microbial electrolysis cell (MEC) with reduced electrode spacing. *Bioresour. Technol.* 102, 3571-3574.Chou, C., Wang, C., Huang, C., Lay, J., (2008). Pilot study of the influence of stirring and pH on anaerobes converting high-solid organic wastes to hydrogen. *Int. J. Hydrogen Energy* 33(5), 1550-1558.

Cheng, X.Y., Li, Q., Liu, C.Z., (2012). Coproduction of hydrogen and methane via anaerobic fermentation of cornstalk waste in continuous stirred tank reactor integrated with up-flow anaerobic sludge bed. *Bioresour. Technol.* 114, 327-333.

Cheng, X.Y., Liu, C.Z., (2010). Enhanced biogas production from herbal-extraction process residues by microwave-assisted alkaline pretreatment. *J. Chem. Technol. Biotechnol.* 85(1), 127-131.

Cheng, X.Y., Liu, C.Z., (2011). Hydrogen production via thermophilic fermentation of cornstalk by Clostridium thermocellum. *Energy Fuels* 25(4), 1714-1720.

Cheng, X.Y., Liu, C.Z., (2012a). Enhanced coproduction of hydrogen and methane from cornstalks by a three-stage anaerobic fermentation process integrated with alkaline hydrolysis. *Bioresour. Technol.* 104, 373-379.

Cheng, X.Y., Liu, C.Z., (2012b). Fungal pretreatment enhances hydrogen production via thermophilic fermentation of cornstalk. *Appl. Energy* 91(1), 1-6.

Cheng, X.Y., Zhong, C., (2014). Effects of Feed to Inoculum ratio, co-digestion and pretreatment on biogas production from anaerobic digestion of cotton stalk. *Energy Fuels* 28(5), 3157-3166.

Chu, F.C., Yu, Y.L., Kai, Q.X., Yoshitaka, E., Yuhei, I., Hai, N.K., (2008). A pH- and temperature-phased two-stage process for hydrogen and methane production from food waste. *Int. J. Hydrogen Energy* 33, 4739-4746.

Das, D., Veziroglu, T.N., (2008). Advances in biological hydrogen production processes. *Int. J. Hydrogen Energy* 33, 6046-6057.

Datar, R., Huang, J., Maness, P.C., Mohagheghi, A., Czernik, S., Chornet, E., (2007). Hydrogen production from the fermentation of corn stover biomass pretreated with a steam-explosion process. *Int. J. Hydrogen Energy* 32(8), 932-939.

Demirbas, A., (2009). Pyrolysis mechanisms of biomass materials. *Energy Source A* 31, 1186-1193.

Demirbas, M.F., Balat, M., Balat, H., (2009). Potential contribution of biomass to the sustainable energy development. *Energy Convers. Manage.* 50, 1746-1760.

Demirel, B., Scherer, P., Yenigun, O., Onay, T.T., (2010). Production of methane and hydrogen from biomass through conventional and high-rate anaerobic digestion processes. *Crit. Rev. Environ. Sci. Technol.* 40, 116-146.

de Vrije, T., Budde, M.A.W., Lips, S.J., Bakker, R.R., Mars, A.E., Claassen, P.A.M., (2010). Hydrogen production from carrot pulp by the extreme thermophiles *Caldicellulosiruptor saccharolyticus* and *Thermotoga neapolitana*. *Int. J. Hydrogen Energy* 35, 13206-13213.

Elsharnouby, O., Hafez, H., Nakhla, G., Naggar, M.H.E., (2013). A critical literature review on biohydrogen production by pure cultures. *Int. J. Hydrogen Energy* 38, 4945-4966.

Eroglu, E., Eroglu, I., Gunduz,U., Yu cel, M., (2009). Treatment of olive mill wastewater by different physicochemical methods and utilization of their liquid effluents for biological hydrogen production. *Biomass Bioenergy* 33(4), 701-705.

Fabiano, B., Perego, P., (2002). Thermodynamic study and optimization of hydrogen production by *Enterobacter aerogenes*. *Int. J. Hydrogen Energy* 27(2), 149-156.

Fang, H.H.P., Liu, H., Zhang, T., (2005). Phototrophic hydrogen production from acetate and butyrate in wastewater. *Int. J. Hydrogen Energy* 30(7), 785-793.

Fan, L.T., Gharpuray, M.M., Lee, Y.H., (1987). Cellulose hydrolysis. Springer-Verlag, Berlin, Germany.

Fan, Y.T., Zhang, G.S,, Guo, X.Y., Xing, Y., Fan, M.H., (2006). Biohydrogen production from beer lees biomass by cow dung compost. *Biomass Bioenergy* 30(5), 493-496.

Fan, Y.T., Xing, Y., Ma, H.C., Pan, C.M., Hou, H.W., (2008). Enhanced cellulose-hydrogen production from corn stalk by lesser panda manure. *Int. J. Hydrogen Energy* 33, 6058-6065.

Fengel, D., Wegener, G., (1984). Wood: Chemistry, Ultrastructure, Reactions. De Gruyter, Berlin.

Fountoulakis, M.S., Manios, T., (2009). Enhanced methane and hydrogen production from municipal solid waste and agro-industrial by-products co-digested with crude glycerol. *Bioresour. Technol.* 100(12), 3043-3047.

Gest, H., Kamen, M.D., (1949). Photoproduction of molecular hydrogen by Rhodospirillum rubrum. *Science* 109, 558-559.

Gioannis, G.D., Muntoni, A., Polettini, A., Pomi, R., (2013). A review of dark fermentative hydrogen production from biodegradable municipal waste fractions. *Waste Manage.* 33, 1345-1361.

Gómez, X., Morán, A., Cuetos, M.J., Sanchez, M.E., (2006). The production of hydrogen by dark fermentation of municipal solid wastes and slaughterhouse waste: a two-phase process. *J. Power Sources* 157, 727-732.

Guo, X.M., Trably, E., Latrille, E., Carrère, H., Steyer, J.P., (2010a). Hydrogen production from agricultural waste by dark fermentation: a review. *Int. J. Hydrogen Energy* 35, 10660-10673.

Guo, K., Tang, X., Du, Z., Li, H., (2010b). Hydrogen production from acetate in a cathode-on-top single-chamber microbial electrolysis cell with a mipor cathode. *Biochem. Eng. J.* 51, 48-52.

Guo, Y.C., Yang, D., Bai, Y.X., Li, Y.H., Fan, Y.T., Hou, H.W., (2014). Co-producing hydrogen and methane from higher concentration of corn stalk by combining hydrogen fermentation and anaerobic digestion. *Int. J. hydrogen energy* 39, 14204-14211.

Hallenbeck, P.C., (2009). Fermentative hydrogen production: Principles, progress, and prognosis. *Int. J. Hydrogen Energy* 34, 7379-7389.

Hamelinck, C.N., Hooijdonk, G., Faaji, A.P.C., (2005). Ethanol from lignocellulosic biomass: techno-economic performance in short-, middle- and long-term. *Biomass Bioenergy* 28(4), 384-410.

Hendriks, A.T.W.M., Zeeman, G., (2009). Pretreatments to enhance the digestibility of lignocellulosic biomass. *Bioresour. Technol.* 100, 10-18.

Hawkes, F.R., Dinsdale, R., Hawkes, D.L., Hussy, I., (2002). Sustainable fermentative hydrogen production: challenges for process optimisation. *Int. J. Hydrogen Energy* 27(11-12), 1339-1347.

Herbert, H.P., Fang, H.L., (2002). Effect of pH on hydrogen production from glucose by a mixed culture. *Bioresour. Technol.* 82, 87-93.

Hillmer, P., Gest, H., (1977). H_2 metabolism in photosynthetic bacterium Rhodopseudomonas capsulata: H_2 production by growing cultures. *J. Bacteriol.* 129(2), 724-731.

Hobson, P.N., Bousfield, S., Summers, R., Kirsch, E.J., (1974). Anaerobic digestion of organic matter. *Crit. Rev. Environ. Sci. Technol.* 4(1), 131-191.

Holm-Nielsen, J., AlSeadi, T., Oleskowicz-Popiel, P., (2009). The future of anaerobic digestion and biogas utilization. *Bioresour. Technol.* 100(22), 5478-5484.

Hong, C., Haiyun, W., (2010). Optimization of volatile fatty acid production with cosubstrate of food wastes and dewatered excess sludge using response surface methodology. *Bioresour. Technol.* 101, 5487-5493.

Hoogwijk, M., Faaij, A., van den Broek, R., Berndes, G., Gielen, D., Turkenburg, W., (2003). Exploration of the ranges of the global potential of biomass for energy. *Biomass Bioenergy* 25(2), 119-123.

Ivanova, G., Rakhely, G., Kovacs, K.L., (2009). Thermophilic biohydrogen production from energy plants by *Caldicellulosiruptor saccharolyticus* and comparison with related studies. *Int. J. Hydrogen Energy* 34(9), 3659-3670.

Jayalakshmi, S., Joseph, K., Sukumaran, V., (2009). Biohydrogen generation from kitchen waste in an inclined plug flow reactor. *Int. J. Hydrogen Energy* 34(21), 8854-8858.

Jeremiasse, A.W., Hamelers, H.V.M., Buisman, C.J.N., (2010). Microbial electrolysis cell with a microbial biocathode. *Bioelectrochemistry* 78, 39-43.

Jonsson, L.J., Alriksson, B., Nilvebrant, N.O., (2013). Bioconversion of lignocellulose: inhibitors and detoxification. *Biotechnol. Biofuels* 6, 16.

Jung, K.W., Kim, D.H., Shin, H.S., (2011). Fermentative hydrogen roduction from Laminaria japonica and optimization of thermal pretreatment conditions. *Bioresour. Technol.* 102(3), 2745-2750.

Kapdan, I.K., Kargi, F., (2006). Bio-hydrogen production from waste materials. *Enzyme Microb. Technol.* 38(5), 569-582.

Karakashev, D., Trably, E., Angelidaki, I., (2009). A strict anaerobic extreme thermophilic hydrogen-producing culture enriched from digested household waste. *J. Appl. Microbiol.* 106(3), 1041-1049.

Karlsson, A., Vallin, L., Ejlertsson, J., (2008). Effects of temperature, hydraulic retention time and hydrogen extraction rate on hydrogen production from the fermentation of food industry residues and manure. *Int. J. Hydrogen Energy* 33(3), 953-962.

Khanal, S.K., Chen, W.H., Li, L., Sung, S., (2004). Biological hydrogen production: effect of pH and intermediate products. *Int. J. Hydrogen Energy* 29, 1123-1131.

Kim, S.H., Han, S.K., Shin, H.S., (2004). Feasibility of biohydrogen production by anaerobic co-digestion of food waste and sewage sludge. *Int. J. Hydrogen Energy* 29(15), 1607-1616.

Kim, S., Han, S., Shin, H., (2006a). Effect of substrate concentration on hydrogen production and 16s rDNA-based analysis of the microbial community in a continuous fermenter. *Process Biochem.* 41(1), 199-207.

Kim, D.H., Han, S.K., Kim, S.H., Shin, H.S., (2006b). Effect of gas sparging on continuous fermentative hydrogen production. *Int. J. Hydrogen Energy* 31(15), 2158-2169.

Kim, D.H., Kim, S.H., Kim, H.W., Kim, M.S., Shin, H.S., (2011). Sewage sludge addition to food waste synergistically enhances hydrogen fermentation performance. *Bioresour. Technol.* 102, 8501-8506.

Kim, D.H., Kim, S.H., Kim, K.Y., Shin, H.S., (2010). Experience of a pilot-scale hydrogenproducing anaerobic sequencing batch reactor (ASBR) treating food waste. *Int. J. Hydrogen Energy* 35, 1590-1594.

Kim, D.H., Kim, S.H., Kim, H.W., Kim, M.S., Shin, H.S., (2011a). Sewage sludge addition to food waste synergistically enhances hydrogen fermentation performance. *Bioresour. Technol.* 102, 8501-8506.

Kim, D.H., Wu, J., Jeong, K.W., Kim, M.S., Shin, H.S., (2011b). Natural inducement of hydrogen from food waste by temperature control. *Int. J. Hydrogen Energy* 36, 10666-10673.

Kim, M.J., Liu, C.G., Noh, J.W., Yang, Y.N., Oh S., Shimizu, K., Lee, D.Y., Zhang, Z.Y., (2013). Hydrogen and methane production from untreated rice straw and raw sewage sludge under thermophilic anaerobic conditions. *Int. J. Hydrogen Energy* 38, 8648-8656.

Kim, S.H., Shin, H.S., (2008a). Effects of base-pretreatment on continuous enriched culture for hydrogen production from food waste. *Int. J. Hydrogen Energy* 33(19), 5266-5274.

Kim, S.H., Han, S.K., Shin, H.S., (2008b). Optimization of continuous hydrogen fermentation of food waste as a function of solids retention time independent of hydraulic retention time. *Process Biochem.* 43, 213-218.

Kirtay, E., (2011). Recent advances in production of hydrogen from biomass *Energy Convers. Manage.* 52, 1778-1789.

Kotay, S.M., Das, D., (2007). Microbial hydrogen production with *Bacillus coagulans* IIT-BT s1 isolated from anaerobic sewage sludge. *Bioresour. Technol.* 98(6), 1183-1190.

Kotsopoulos, T.A., (2009). Biohydrogen production from pig slurry in a CSTR reactor system with mixed cultures under hyperthermophilic temperature (70 °C). *Biomass Bioenergy* 33(9), 1168-1174.

Koutrouli, E.C., Kalfas, H., Gavala, H.N., Skiadas, I.V., Stamatelatou, K., Lyberatos, G., (2009). Hydrogen and methane production through two-stage mesophilic anaerobic digestion of olive pulp. *Bioresour. Technol.* 100(15), 3718-3723.

Kuyper, M., Toirkens, M.J., Diderich, J.A., Winkler, A.A., Van Dijken, J.P., Pronk, J.T., (2005). Evolutionary engineering of mixed-sugar utilization by a xylose-fermenting *Saccharomyces cerevisiae* strain .*FEMSYeast Res.* 5 (10), 925-934.

Kongjan, P., O-Thong, S., Angelidaki, I., (2011). Performance and microbial community analysis of two-stage process with extreme thermophilic hydrogen and thermophilic methane production from hydrolysate in UASB reactors. *Bioresour. Technol.* 102, 4028-4035.

Kumar, N., Das, D., (2000). Enhancement of hydrogen production by Enhancement of hydrogen production by *Enterobacter cloacae* IIT-BT 08. *Process Biochem.* 35, 6, 589-593.

Lakaniemi, A.M., Hulatt, C.J., Thomas, D.N., Tuovinen, O.H., Puhakka, J.A., (2011). Biogenic hydrogen and methane production from *Chlorella vulgaris* and *Dunaliella tertiolecta* biomass. *Biotechnol. Biofuels* 4, 34.

Lateef, S.A., Beneragama, N., Yamashiro, T., Iwasaki, M., Umetsu, K., (2014). Batch anaerobic co-digestion of cow manure and waste milk in two-stage process for hydrogen and methane productions. *Bioprocess Biosystems Eng.* 37(3), 355-363.

Lay, J.J., Fan, K.S., Chang, J., Ku, C.H., (2003). Influence of chemical nature of organic wastes on their conversion to hydrogen by heat-shock digested sludge. *Int. J. Hydrogen Energy* 28, 1361-1367.

Lalaurette, E., Thammannagowda, S., Mohagheghi, A., Maness, P.-C., Logan, B.E., (2009). Hydrogen production from cellulose in a two-stage process combining fermentation and electrohydrogenesis. *Int. J. Hydrogen Energy* 34, 6201-6210.

Laureano-Perez, L., Teymouri, F., Alizadeh, H., Dale, B.E., (2005). Understanding factors that limit enzymatic hydrolysis of biomass. *Appl. Biochem. Biotechnol.* 1081-1099.

Lee, W.G., Lee, J.S., Shin, C.S., Park, S.C., Chang, H.N., Chang, Y.K., (1999). Ethanol production using concentrated oak wood hydrolysates and methods to detoxify. *Appl. Biochem. Biotechnol.* 77-79, 547-549.

Lee, Y.W., Chung, J., (2010). Bioproduction of hydrogen from food waste by pilot-scale combined hydrogen/methane fermentation. *Int. J. Hydrogen Energy* 35, 11746-11755.

Lee, Z.K., Li, S.L., Kuo, P.C., Chen, I.C., Tien, Y.M., Huang, Y.J., Chuang, C.P., Wong, S.C., Cheng, S.S., (2010). Thermophilic bio-energy process study on hydrogen fermentation with vegetable kitchen waste. *Int. J. Hydrogen Energy* 35, 13458-13466.

Lee, Z.K., Li, S.L., Lin, J.S., Wang, Y.H., Kuo, P.C., Cheng, S.S., (2008). Effect of pH in fermentation of vegetable kitchen wastes on hydrogen production under a thermophilic condition. *Int. J. Hydrogen Energy* 33, 5234-5241.

Levin, D.B., Chahine, R., (2010). Challenges for renewable hydrogen production from biomass. *Int. J. Hydrogen Energy* 35, 4962-4969.

Levin, D.B., Pitt, L., Love, M., (2004). Biohydrogen production: prospects and limitations to practical application. *Int. J. Hydrogen Energy* 29, 173-185.

Li, C., Fang, H.H.P., (2007). Fermentative hydrogen production from wastewater and solid wastes by mixed cultures. *Crit. Rev. Environ. Sci. Technol.* 37, 1-39.

Li, D., (2009). Hydrogen production characteristics of the organic fraction of municipal solid wastes by anaerobic mixed culture fermentation. *Int. J. Hydrogen Energy* 34(2), 812-820.

Li, D.M., Chen, H.Z., (2007). Biological hydrogen production from steamexploded straw by simultaneous saccharification and fermentation. *Int. J. Hydrogen Energy* 32(12), 1742-1748.

Li, H.Z., Guo, S., Cui, L.Y., Yan, J.J., Liu, J.J., Wang, B., (2015). Review of renewable energy industry in Beijing: Development status, obstacles and proposals. *Renew. Sustain. Energy Rev.* 43, 711-725.

Li, M., Zhao, Y., Guo, Q., Qian, X., Niu, D., (2008a). Bio-hydrogen production from food waste and sewage sludge in the presence of aged refuse excavated from refuse landfill. *Renew Energy* 33(12), 2573-2579.

Li, S.L., Kuo, S.C., Lin, J.S., Lee, Z.K., Wang, Y.H., Cheng, S.S., (2008b). Process performance evaluation of intermittent–continuous stirred tank reactor for anaerobic hydrogen fermentation with kitchen waste. *Int. J. Hydrogen Energy* 33, 1522-1531.

Lin, C.Y., Chang, R.C., (1999). Hydrogen production during the anaerobic acidogenic conversion of glucose. *J. Chem. Technol. Biotechnol.* 74(6), 498-500.

Lin, C.Y., Lay, C.H., (2004). Carbon/nitrogen ratio effect on fermentative hydrogen production by mixed microflora. *Int. J. Hydrogen Energy* 29, 41-45.

Lin, Y.Q., Wu, S.B., Wang, D.H., (2013a). Hydrogen-methane production from pulp & paper sludge and food waste by mesophilicethermophilic anaerobic co-digestion. *Int. J. Hydrogen Energy* 38, 15055-15062.

Lin, Y.Q., Liang, J.J., Wu, S.B., Wang, B.H., (2013b). Was pretreatment beneficial for more biogas in any process? Chemical pretreatment effect on hydrogen-methane co-production in a two-stage process. *J. Industr. Eng. Chem.* 19, 316-321.

Li, Q., Guo, C., Liu, C.Z., (2014). Dynamic microwave-assisted alkali pretreatment of cornstalk to enhance hydrogen production via co-culture fermentation of *Clostridium thermocellum* and *Clostridium thermosaccharolyticum*. *Biomass Bioenergy* 64, 220-229.

Liu, H., Grot, S., Logan, B.E., (2005). Electrochemically assisted microbial production of hydrogen from acetate. *Environ. Sci. Technol.* 39, 4317-4320.

Liu, X.Y., Li, R.Y., Ji, M., Han, L., (2013). Hydrogen and methane production by co-digestion of waste activated sludge and food waste in the two-stage fermentation process: Substrate conversion and energy yield. *Bioresour. Technol.* 146, 317-323.

Li, Q., Liu, C.Z., (2012). Co-culture of Clostridium thermocellum and *Clostridium thermosaccharolyticum* for enhancing hydrogen production via thermophilic fermentation of cornstalk waste. *Int. J. Hydrogen Energy* 37, 10648-10654.

Liu, C.Z., Cheng, X.Y., (2009). Microwave-assisted acid pretreatment for enhancing biogas production from herbal-extraction process residue. *Energy Fuels* 23(12), 6152-6155.

Liu, C.Z., Cheng, X.Y., (2010). Improved hydrogen production via thermophilic fermentation of corn stover by microwave-assisted acid pretreatment. *Int. J. Hydrogen Energy* 35(17), 8945-8952.

Liu, D.W., Liu, D.P., Zeng, R.J., Angelidaki, I., (2006). Hydrogen and methane production from household solid waste in the two-stage fermentation process. *Wat. Res.* 40, 2230-2236.

Liu, Z.D., Lv, F.X., Zheng, H., Zhang, C., Wei, F., Xing, X.H., (2012). Enhanced hydrogen production in an UASB reactor by retaining microbial consortium onto carbon nanotubes (CNTs). *Int. J. Hydrogen Energy.* 37, 10619 -10626.

Liu, Z.D., Liu, J., Zhang, S.P., Xing, X.H., Su, Z.G., (2011). Microbial fuel cell based biosensor for in situ monitoring of anaerobic digestion process. *Bioresour. Technol.* 102, 10221-10229.

Li, Y., Sasaki, H., Torii, H., Okuno, Y., Seki, K., Kamigochi, I., (1999). Comparison between mesophilic and thermophilic high solids anaerobic digestion in treating the organic fraction of MSW. *Environ. Eng. Res.* 346-354.

Logan, B.E., Call, D., Cheng, S., Hamelers, H.V.M., Sleutels, T.H.J.A., Jeremiasse, A.W., Rozendal, R.A., (2008). Microbial electrolysis cells for high yield hydrogen gas production from organic matter. *Environ. Sci. Technol.* 42, 8630-8640.

Lü, F., Ji, J.Q., Shao, L.M., He, P.J., (2013). Bacterial bioaugmentation for improving methane and hydrogen production from microalgae. *Biotechnol. Biofuels* 6, 92

Lu, Y., Lai, Q.H., Zhang, C., Zhao, H.X., Ma, K., Zhao, X.B., Chen, H.Z., Liu, D.H., Xing, X.H., (2009). Characteristics of hydrogen and methane production from cornstalks by an augmented two- or three-stage anaerobic fermentation process. *Bioresour. Technol.* 100, 2889-2895.

Magnusson, L., Islam, R., Sparling, R., Levin, D., Cicek, N., (2008). Direct hydrogen production from cellulosic waste materials with a single-step dark fermentation process. *Int. J. Hydrogen Energy* 33, 5398-5403.

Mandal, B., Nath, K., Das, D., (2006). Improvement of biohydrogen production under decreased partial pressure of H_2 by *Enterobacter cloacae. Biotechnol. Lett.* 28(11), 831-835.

Mosier, N., Wyman, C., Dale, B., Elander, R., Lee, Y.Y., Holtzapple, M., Ladisch, M., (2005b). Features of promising technologies for pretreatment of lignocellulosic biomass. *Bioresour. Technol.* 96, 673-686.

Mtui, G.Y.S., (2009). Recent advances in pretreatment of lignocellulosic wastes and production of value added products. *Afr. J. Biotechnol.* 8(8), 1398-1415.

Nath, K., Das, D., (2003). Hydrogen from biomass. *Curr. Sci.* 85(3), 265-271.

Nath, K., Muthukumar, M., Kumar, A., Das, D., (2008). Kinetics of two-stage fermentation process for the production of hydrogen. *Int. J. Hydrogen Energy* 33(4), 1195-1203.

Nazlina, H.M.Y.N.H., NorAini, A.R., Man, H.C., Yusoff, M.Z.M., Hassan, M.A., (2011). Microbial characterization of hydrogen-producing bacteria in fermented food waste at different pH values. *Int. J. Hydrogen Energy* 36, 9571-9580.

Ngo, T.A., Nguyen, T.H., Bui, H.T.V., (2012). Thermophilic fermentative hydrogen production from xylose by *Thermotoga neapolitana* DSM 4359. *Renew. Energy* 37, 174-179.

Nissila, M.E., Lay, C.H., Puhakka, J.A., (2014). Dark fermentative hydrogen production from lignocellulosic hydrolyzatese A review. *Biomass Bioenergy* 67, 145-159.

Nissila, M.E., Li, Y.C., Wu, S.Y., Lin, C.Y., Puhakka, J.A., (2012). Hydrogenic and methanogenic fermentation of birch and conifer pulps. *Appl. Energy* 100, 58-65.

Okamoto, M., Noike, T., Miyahara, T., Mizuno, O., (2000). Biological hydrogen potential of materials characteristic of the organic fraction of municipal solid wastes. *Water Sci. Technol.* 41(3), 25-32.

O-Thong, S., Prasertsan, P., Intrasungkha, N., Dhamwichukorn, S., Birkeland, N., (2008a). Optimization of simultaneous thermophilic fermentative hydrogen production and COD reduction from palm oil mill effluent by thermoanaerobacterium-rich sludge. *Int. J. Hydrogen Energy* 33(4), 1221-1231.

O-Thong, S., Prasertsan, P., Karakashev, D., Angelidaki, I., (2008b). Thermophilic fermentative hydrogen production by the newly isolated *Thermoanaerobacterium thermosaccharolyticum* PSU-2. *Int. J. Hydrogen Energy* 33, 1204-1211.

O-Thong, S., Prasertsan, P., Karakashev, D., Angelidaki, I., (2008c). 16srRNA-targeted probes for specific detection of *Thermoanaerobacterium*spp.,*Thermoanaerobacterium thermosaccharolyticum*, and *Caldicellulosiruptor* spp. By fluorescent in situ hybridization in biohydrogen producing systems. *Int. J. Hydrogen Energy* 33(21), 6082-6091.

Pakarinen, O., Lehtomaki, A., Rintala, J., (2008). Batch dark fermentative hydrogen production from grass silage: the effect of inoculum, pH, temperature and VS ratio. *Int. J. Hydrogen Energy* 33(2), 594-601.

Pan, J., Zhang, R., El-Mashad, H.M., Sun, H., Ying, Y., (2008). Effect of food to microorganism ratio on biohydrogen production from food waste via anaerobic fermentation. *Int. J. Hydrogen Energy* 33, 6968-6975.

Pan, C., Zhang, S., Fan, Y., Hou, H., (2010). Bioconversion of corncob to hydrogen using anaerobic mixed microflora. *Int. J. Hydrogen Energy* 35(7), 2663-2669.

Park, J.I., Lee, J., Sim, S.J., Lee, J.H., (2009). Production of hydrogen from marine macroalgaebiomass using anaerobic sewage sludge microflora. *Biotechnol. Bioprocess Eng.* 14, 307-315.

Pattra, S., Sangyoka, S., Boonmee, M., Reungsang, A., (2008). Biohydrogen production from the fermentation of sugarcane bagasse hydrolysate by *Clostridium butyricum*. *Int. J. Hydrogen Energy* 33, 5256-5265.

Quemeneur, M., Bittel, M., Trably, E., Dumas, C., Fourage, L., Ravot, G., Steyera, J.P., Carrèrea, H., (2012). Effect of enzyme addition on fermentative hydrogen production from wheat straw. *Int. J. Hydrogen Energy* 37(14), 10639-10647.

Rabelo, S.C., Carrere, H., Maciel Filho, R., Costa, A.C., (2011). Production of bioethanol, methane and heat from sugarcane bagasse in a biorefinery concept. *Bioresour. Technol.* 102: 7887-7895.

Rani, K.S., Swamy, M.V., Seenayya, G., (1998). Production of ethanol from various pure and natural cellulosic biomass by *Clostridium thermocellum* strains SS21 and SS22. *Process Biochem.* 33: 435-440.

Ren, N.Q., Wang, A.J., Cao, G., Xu, J., Gao, L., (2009). Bioconversion of lignocellulosic biomass to hydrogen: potential and challenges. *Biotechnol. Adv.* 27(6), 1051-1060.

Sainio, T., Turku, I., Heinonen, J., (2011). Adsorptive removal of fermentation inhibitors from concentrated acid hydrolysates of lignocellulosic biomass. Bioresour. Technol.;102(10), 6048-6057.

Salerno, M.B., Park, W., Zuo, Y., Logan, B.E., (2006). Inhibition of biohydrogen production by ammonia. *Water Res.*40(6), 1167-1172.

Saratale, G.D., Chen, S.D., Lo, Y.C., Saratale, R.G., Chang, J.S., (2008). Outlook of biohydrogen production from lignocellulosic feedstock using dark fermentation e a review. *J. Sci. Ind. Res.* 67, 962-979.

Shen, J.F., Luo, C., (2015). Overall review of renewable energy subsidy policies in China - Contradictions of intentions and effects. *Renew. Sustain. Energy Rev.* 41, 1478-1488.

Shin, H., Youn, J., (2005). Conversion of food waste into hydrogen by thermophilic acidogenesis. *Biodegradation* 16(1), 33-44.

Shin, H.S., Youn, J.H., Kim, S.H., (2004). Hydrogen production from food waste in anaerobic mesophilic and thermophilic acidogenesis. *Int. J. Hydrogen Energy* 29, 1355-1363.

Srirangan, K., Akawi, L., Moo-Young, M., Chou, C.P., (2012). Towards sustainable production of clean energy carriers from biomass resources. *Appl. Energy* 100, 172-186.

Taguchi, F., Mizukami, N., Taki, T.S., Hasegawa, K., (1995). Hydrogen-production from continuous fermentation of xylose during growth of *Clostridium* sp strain No-2. *Can. J. Microbiol.* 41, 536-540.

Tanisho S. Feasibility study of biological hydrogen production from sugar cane by fermentation. In: Veziroglu TN, Winter C-J, Basselt JP, Kreysa G, eds. Hydrogen Energy Progress XI. Proceedings of the 11th WHEC. Stuttgart, vol. 3; 1996. p. 2601-2606.

Tanisho, S., Ishiwata, Y., (1994). Continuous hydrogen-production from molasses by the bacterium *Enterobacter aerogenes*. *Int. J. Hydrogen Energy* 19, 807-812.

Tanisho, S., Ishiwata, Y., (1995). Continuous hydrogen-production from molasses by fermentation using urethane forms as a support of flocks. *Int. J. Hydrogen Energy* 20, 541-545.

Tao, Y., Hea, Y., Wu, Y., Liu, F., Li, X., Zong, W., Zhou, Z.H., (2008). Characteristics of a new photosynthetic bacterial strain for hydrogen production and its application in wastewater treatment. *Int. J. Hydrogen Energy* 33, 963-973.

Tuna, E., Kargi, F., Argun, H., (2009). Hydrogen gas production by electrohydrolysis of volatile fatty acid (VFA) containing dark fermentation effluent. *Int. J. Hydrogen Energy* 34, 262-269.

Tuncel, F., Gercel, H.F., (2004). Production and characterization of pyrolysis oils from Euphorbia macroclada. *Energy Source A* 26, 761-770.

Wang, A., Sun, D., Cao, G., Wang, H., Ren, N., Wu, W.-M., Logan, B.E., (2011). Integrated hydrogen production process from cellulose by combining dark fermentation, microbial fuel cells, and a microbial electrolysis cell. *Bioresour. Technol.* 102, 4137-4143.

Wang, X., Zhao, Y., (2009). A bench scale study of fermentative hydrogen and methane production from food waste in integrated two-stage process. *Int. J. Hydrogen Energy* 34, 245-254.

Weiland, P., (2006). Biomass digestion in agriculture: a successful pathway for the energy production and waste treatment in Germany. *Eng. Life Sci.* 6(3), 302-309.

van Niel, E.W.J., Budde, M.A.W., de Haas, G.G., van der Wal, F.J., Claasen, P.A.M., Stams, A.J.M., (2002). Distinctive properties of high hydrogen producing extreme thermophiles, *Caldicellulosiruptor saccharolyticus* and *Thermotoga elfii*. *Int. J. Hydrogen Energy* 27, 1391-1398.

Valdez-Vazquez, I., Rios-Leal, E., Esparza-García, F., Cecchi, F., Poggi-Varaldo, H., (2005). Semi-continuous solid substrate anaerobic reactors for H_2 production from organic waste: mesophilic versus thermophilic regime. *Int. J. Hydrogen Energy* 30, 1383-1391.

Xing,Y., Li, Z., Fan, Y.T., Hou, H.W., (2010). Biohydrogen production from dairy manures with acidification pretreatment by anaerobic fermentation. *Environ. Sci. Pollut. Res.* 17, 392-399.

Yokoi, H., Maki, R., Hirose, J., Hayashi, S., (2002). Microbial production of hydrogen from starch-manufacturing wastes. *Biomass Bioenergy* 22(5), 389-395.

Yokoi, H., Maeda, Y., Hirose, J., Hayashi, S., Takasaki, Y., (1997). H_2 production by immobilized cells of *Clostridium butyricum* on porous glass beads. *Biotechnol. Tech.* 11, 431-433.

Yang, Z.M., Guo, R.B., Xu, X.H., Fan, X.L., Luo, S.J., (2011). Fermentative hydrogen production from lipid-extracted microalgal biomass residues. *Appl. Energy* 88, 3468-3472.

Yokoyama, H., Waki, M., Moriya, N., Yasuda, T., Tanaka, Y., Haga, K., (2007). Effect of fermentation temperature on hydrogen production from cow waste slurry by using anaerobic microflora within the slurry. *Appl. Microbiol. Biotechnol.* 74(2), 474-483.

Zeng, M., Mosier, N.S., Huang, C.P., Sherman, D.M., Ladisch, M.R., (2007). Microscopic examination of changes of plant cell structure in corn stover due to cellulase activity and hot water pretreatment. *Biotechnol. Bioeng.* 97(2).

Zhang, M.L., Fan, Y.T., Xing, Y., Pan, C.M., Zhang, G.S., Lay, J.J., (2007). Enhanced biohydrogen production from Corn stalk wastes with acidification pretreatment by mixed anaerobic cultures. *Biomass Bioenergy* 31(4), 250-254.

Zhu, H., Parker, W., Basnar, R., Proracki, A., Falletta, P., Beland, M., Seto, P., (2008). Biohydrogen production by anaerobic co-digestion of municipal food waste and sewage sludges. *Int. J. Hydrogen Energy* 33, 3651-3659.

Zhu, J., Miller, C., Li, Y.C., Wu, X., (2009). Swine manure fermentation to produce biohydrogen. *Bioresour. Technol.* 100(22), 5472-5477.

Zong, W., Yu, R., Zhang, P., Fan, M., Zhou, Z., (2009). Efficient hydrogen gas production from cassava and food waste by a two-step process of dark fermentation and photo-fermentation. *Biomass Bioenergy* 33, 1458-1463.

In: Gas Biofuels from Waste Biomass
Editor: Zhidan Liu

ISBN: 978-1-63483-192-5
© 2015 Nova Science Publishers, Inc.

Chapter 9

THE CULTIVATION OF MICROALGAE BY USING ANIMAL AND POULTRY DIGESTION EFFLUENT TO REALIZE USEFUL BIOMASS PRODUCTION AND NUTRIENTS RECYCLING

Haifeng Lu[*] and Guoyang Yuan

Collecge of Water Resource and Civil Engineering,
China Agriculture University, Beijing, China

ABSTRACT

Disposal of animal and poultry digestion effluent often results in high nutrient loading into aquatic environments, which may lead to favorable conditions for undesirable phytoplankton blooms. Microalgae are efficient in removing nitrogen, phosphorus and toxic metals from wastewater under controlled environments. If key nutrients in the wastewater stream can be used to grow microalgae for fertilizer or biofuel production, the nutrients can be removed, thus the risk of the over growth of harmful phytoplankton could be significantly reduced. This portion summarizes the major nutrient components of different animal and poultry digestion effluent streams, the mechanisms of algal nutrient uptake, the studies on microalgae cultivation by different animal and poultry digestion effluent, and current microalgae production systems. Finally, some suggestions were proposed.

Keywords: microalage cultivation, animal and poultry digestion effluent treatment, microalgae production system

[*] Corresponding Author address: Qinghuadonglu 17. Mail box, No. 67, Haidian District, Beijing, China, 100083, Email: hfcauedu@163.com

1. INTRODUCTION

Microalgae have been proven to be efficient in removing nitrogen, phosphorus and toxic metals from a wide variety of animal and poultry digestion effluent (Zhou et al., 2012; Larson, 1979; Lindstrom et al., 1981). There are extensive studies of algae growth in swine, dairy, cattle, poultry manure anaerobic digestion effluent (Ragauskas et al., 2006; Sheehan et al., 1998). Substantial amounts of nutrient removal, hydrogen production and algae biomass production were obtained in these studies. Hence, controlled microalgae cultivation shows promises a potential biological treatment method for animal and poultry digestion effluent.

2. THE MICROALGAE CULTIVATION IN THESE STREAM

2.1. The Mechanism of Pollutants Removal

In order to maximize nutrient removal from wastewater streams by various species of microalgae, the mechanisms of algal nutrient uptake should be thoroughly understood. The various metabolic pathways of the algal cell can be coarsely categorized by elemental constituent. In addition to the four basic elements, i.e., carbon, nitrogen, phosphorus and sulfur, ionic components such as sodium, potassium, iron, magnesium, calcium, and trace elements must be provided for algal growth as well. Emphasis is usually put on nitrogen and phosphorus. Anthropogenic waste production can significantly increase these nutrients which can lead to a risk of nutrient run off and eutrophication.

2.1.1. Carbon

Generally, carbon can be fixed through the photosynthetic activity of autotrophic microalgae in the form of carbon dioxide. Some microalgae display heterotrophic or/and mixotrophic behavior, using organic forms of carbon to form the carbon skeleton of their body. Carbon can be also utilized in the form of soluble carbonates for cell growth, either by direct uptake or conversion of carbonate to free carbon dioxide through carboanhydrase activity. The use of algae to mitigate carbon dioxide from flue gases is another research focus, and if it is effective, both the environment and biofuel production could benefit from that.

2.1.2. Nitrogen

Nitrogen is a critical nutrient required in the growth of all organisms. In microalgae, nitrogen mainly existed in the form of organic nitrogen. It is found in avariety of biological substances, such as peptides, proteins, enzymes, chlorophylls, energy transfer molecules (ADP, ATP), and genetic materials (RNA, DNA) (Barsanti et al., 2006). It is derived from inorganic source sincluding nitrate (NO_3^-), nitrite (NO_2^-), nitricacid (HNO_3), ammonium (NH_4^+), ammonia (NH_3), and nitrogengas (N_2). There are two kinds of inorganic nitrogen transformation, assimilation and fixation. Assimilation occurred mostly in alleukaryoticalgae, through assimilation process, most microalgae convert in organic nitrogen to its organic form. For assimilation, the nitrogen must be in the forms of nitrate, nitrite or ammonium. While for cyanobacteria, they also can convert atmospheric nitrogen into ammonia by means of fixation.

For the assimilation of inorganic nitrogen, the process can be described as follows: Firstly, translocation of the inorganic nitrogen occurs across the plasma membrane, followed by the reduction of oxidized nitrogen and the incorporation of ammonium into aminoacids. NO_3^- and NO_2^- undergo reduction with the assistance of nitrate reductase and nitrite reductase, respectively. Nitrate reductase uses the reduced form of nicotin amidea denine dinucleotide (NADH) to transfer two electrons, conversing NO_3^- into NO_2^-. NO_2^- is reduced to NH_4^+-N by nitrite reductase and ferre doxin (Fd), transferring a total of six electrons in the reaction. All forms of inorganic nitrogen are ultimately reduced to NH_4^+-N prior to being incorporate into aminoacids within the intracellular fluid. Finally, glutamate (Glu), adenosine triphosphate (ATP) and glutamine synthase facilitates the NH_4^+-N transport into the aminoacid glutamine.

Generally, microalgae prefer NH_4^+-N over NO_3^-, and NO_3^- absorption does not occur until the NH_4^+-N in the environment is almost completely consumed (Maestrini et al., 1986). That might be caused by the uninvolved assimilation of the redox reaction. The absorption of NH_4^+-N requires less energy. Therefore, wastewater with high NH_4^+-N concentrations can be effectively used to rapidly grow microalgae. However, NO_3^- can also be an essential nitrogen source for microalgae as the presence of NO_3^- induces the activity of nitrate reductase. Incontrast, excess NH_4^+-N can have a repressive effect (Morris and Syrett, 1963). The NH_4^+-N tolerance of different microalgae species varies from 25 mmol/L NH_4^+-N to 1000 mmol/L NH_4^+-N (Collos and Berges, 2004). One approach to growing microalgae in high NH_4^+-N concentration wastewater is utilizing the plant enzyme glutamine synthetase, which has a high affinity for NH_4^+-N. The addition of glutamic acid was reported to result in 70% more NH_4^+-N reduction percell during the growth of *Chlorella vulgaris* in natural wastewater (Khan and Yoshida, 2008).

Except cell metabolism, NH_4^+-N also can be removed by stripping in wastewater treatment duration. The factors that influenced the stripping is pH and temperature. The higher the pH and temperature are, the easier the stripping efficiency will be achieved. In fact, NH_4^+-N stripping was the most important mechanism in high growth rate algal pond so perating at various hydraulic retention times (Garcia et al., 2000). When the algal pond was exposed to warm weather, ammonia release accelerated even when the pH was below 9.

2.1.3. Phosphorus

Phosphorus is also a key factor in the energy metabolism of microalgae. It is also a very important element in nucleic acids, lipids, proteins and the intermediates of carbohydrate metabolism. Inorganic phosphates play a significant role in algae cell growth and metabolism. $H_2PO_4^-$ and HPO_4^{2-} are two main forms that incorporated into organic compounds through phosphorylation to generate ATP from adenosine diphosphate (ADP), accompanied by a form of energy input (Martinez et al., 1999). Energy input can come from the oxidation of organic substrates during the respiration processing, the electron transport system of the mitochondria, or from light through photosynthesis processing. The inorganic forms of phosphorus could be used not only by microalgae, but also by varieties of algae as inorganic ester for growth (Kuenzler, 1965). Although orthophosphate is generally recognized as the limiting nutrient in fresh water systems, many cases of eutrophication are triggered by superfluous phosphorus, which can result from run off of wastewater (Correll, 1998). Phosphorus removal in wastewater is not only governed by the uptake into the cell, but also by external conditions such as pH and dissolved oxygen.

2.2. Anaerobic Digestion Stream and the Characteristics

In the past decades, many efforts were made to investigate the nutrients uptake and biomass production by using animal and poultry anaerobic digestion effluent for environment purification, nutrients utilization and useful biomass production. The composition was various according to different feedstock and fermentation conditions from different fermentation project, which lead to the different results of nutrients removal efficiency and biomass production. In this section, the characteristics of anaerobic digestion stream from different feedstock were summarized. The specific details can be seen in Table 1.

There are mainly four kinds of anaerobic digestion stream, dairy, poultry, swine and beef feedlot. Comparing to the industrial and municipal wastewater, the carbon contents in anaerobic digestion stream form animal and poultry manure are relatively low, and the nitrogen and phosphorus are relatively high. That because microbial activity during the digestion converts the carbon to methane. In animal and poultry anaerobic digestion stream, NH_4^+-N was the mainly form of total nitrogen. Although the TN and TP are very different from each other, the highest N/P levels appeared in poultry anaerobic digestion stream, which might be caused by the initial feedstock materials.

Table 1. Total nitrogen and total phosphorus content of different animal and poultry waste stream (Cai et al., 2013)

Wastewater category	TN (mg/L)	TP (mg/L)	N/P	References
Dairy	185-2636	3-727	3.6-7.2	Barker et al., 2001; Bradford et al., 2008;
Poultry	802-1825	50-446	4-16	Barker et al., 2001; Yetilmezsoy et al., 2008
Swine	1110-3213	310-987	3.0-7.8	Millmier et al., 2000; Barker et al., 2001;
Beef feedlot	63-4165	14-1195	2.0-4.5	Barker et al., 2001; Yetilmezsoy et al., 2008

2.3. The Cultivation of Microalgae Using AD

There are many studies about using the AD effluent from four different feedstocks for microalgae cultivation. The main and typical studies are summarized as follows.

There are more studies concentrated on using swine and dairy anaerobic digestion stream to cultivate microalgae, and fewer studies concentrated on that of using poultry anaerobic digestion stream. That might be caused by the characteristics of the wastewater. Although microalgae have strong ability to consume nitrogen and phosphorus, the N/P in poultry anaerobic digestion stream might be higher than other anaerobic digestion stream, which leading to the lack phosphorus elements.

Swine anaerobic digestion stream used for microalgae cultivation

Table 2. Typical studies about using swing anaerobic digestion stream to cultivate microalgae

Microalgae strain	Purpose	Wastewater concentration (mg/L)				Pretreatment methods	Metabolic mode	Cultivation mode	Biomass production	Nutrients removal	References
		TOC	TN	NH_4^+-N	TP						
Scenedesmus obliquus; Chlorella sorokiniana; Spirulinaplatensis; Euglenaviridis	Pollutants removal	-	72-152	-	6-9	Remove suspended solid trough centrifugation; diluted	Photosynthetic oxygenation	Batch	-	TOC: 42%-55%; NH_4^+-N: 21%-39%	Godos et al., 2010
Scenedesmussp.	Biomass production	-	-	210	-	Remove suspended solid trough centrifugation; diluted	Auto phototrophic	Bacth	0.02 g/L/d	TN: 57.7%; TP: 45.5%	Bjornsson et al., 2013
Chlorellasp.	Biomass production and nutrients removal	COD 689.6-3448	81.2-406.2	64.1-320.4	16.3-81.4	Heated to kill the methanogenicbacteria	Mixotrophic	Bench-scale	0.003-0.07 g/L/d	COD: 751.33 mg/L/d; TP: 20.21 mg/L/d; TN: 38.35 mg/L/d; NH_4^+-N: 60.39 mg/L/d;	Hu et al., 2013
Chlorella vulgaris	Biomass production	COD 4755-5634	173-230	121-154	118-145	Diluted	Mixotrophic	Laboratory-scale model; full-scale model consisted of a tertiary pond	177.3-292.7 mg/L/d	COD: 14%-57%; NH_4^+-N: 33.1%-79.2%; TP: 10.6%-74.0%	Travieso et al., 2004

Table 2. (Continued)

Microalgae strain	Purpose	Wastewater concentration (mg/L)				Pretreatment methods	Metabolic mode	Cultivation mode	Biomass production	Nutrients removal	References
		TOC	TN	NH$_4^+$-N	TP						
Chlorella sp.	Biofuel feedstock production and nutrient removal	-	-	78-180	-	Diluted	Mixotrophic	laboratory-scale model, batch	0.9 d^{-1}	COD: 58.0-61.7%; NH$_3$-N: 31.6-44.7%; TN: 22.3-33.6%; TP: 23.4-70.9%	Hu et al., 2012
Scenedesmus intermedius Chod. and *Nannochloris* sp.	Biomass growth and nutrient removal	-	-	-	-	Filtrated	-	Laboratory experiments; immobilized and free cell cultures	0.014-0.018 d^{-1}	TN: 0.009-0.014 mg/h^{-1}; TP: 0.006-0.022 mg/h^{-1}	Jiménez-Pérez et al., 2004
Scenedesaceae sp.	Biomass growth and nutrients removal	COD 223.5-1117.6	-	64.6-322.9	1.3-6.4	Filtrated and diluted	Heterotrophic	Lab scale batch	0.2g/L/d	COD: 34.4%-37.1%;TN:>96%; TP: 75.7%-80.4%	Cheng and Tian, 2013
Chlorella vulgaris	Biomass growth and production	-	-	10-60	-	Filtrated and diluted	-	Lab scale batch	5.80-10.51 million/ml	NH$_4^+$-N: 41.3%-61.8%	Kumar et al., 2010
Chlamydomonas; Ankistrodesmus; Protoderma; Selenastrum; Chlorella; Oocystis; Achnanthes; Nitzschia; Microspora;	Pollutants removal	COD 526-4326;	59-370	-	-			Long-term operation; two identicalraceway	21-28 mg/m^2/d	COD: 76%; KTN: 88%	Godos et al., 2009
Oocystis sp. and *Scenedesmus* sp.	Degrade pollutants in swine slurry	COD 3858	-	1664	-		Mixotrophic	Open pond and a tubular enclosed photobioreactor	0.332/0.007	COD: 57.6%-67.2%;NH$_4^+$-N: 80.1%-99.9%; TP: 54.3%-84.1%	Molinuevo-Salces et al., 2010
Chlorella sorokiniana	nutrients and carbon removal	1247		656	117	Centrifuged	-	Continuous experiment with tubularbiofilmphotobioreactor		TN: 94-100%; TP: 70-90%	Godos et al., 2009

Purpose	Wastewater concentration (mg/L) Pretreatment methods				Metabolic mode	Cultivation mode	Biomass production	Nutrients removal	References	Purpose
f	TOC	TN	NH$_4^+$-N	TP						
Chlorella sp. Fuel production and wastewater treatment	370-493	245-327	125.0-166.7			Mixotrophoic	Continuous experiment with a greenhouse-based vertically arranged multilayer structure	8.08 to 14.59 and 19.15-23.19 g/m²/d	The NH$_3$-N: 2.65; TN: 3.19; COD: 7.21; TP: 0.067 g/m²/d	Min et al., 2014
Rhizoclonium hieroglyphicum Fatty acid content and composition of algal biomass		0.2-1.3 g/m²/d				Auto phototrophic	Indoor laboratory-scale algal turf scrubber; semi-continuous mode	6.8-10.7 g/m²/d		Mulbry et al., 2008
Spirulina Arthrospira Biomass production and nutrient removal						Mixotrophic	Outdoor raceways on pilot plant; Batch or semi-continuous	0.15 m depth: 14.4 g/m²/d; 0.20 m depth: 15.1 g/m²/d; 0.10 m depth: 11.8 g/m²/d;		Olguín et al., 2003
Botryococcus braunii Hydrocarbon production		10-1012		3-285		Heterotrophic mode		0.014-0.034 d⁻¹		An et al., 2003

Dairy anaerobic digestion stream used for microalgae cultivation

Table 3. Typical studies about using dairy anaerobic digestion stream to cultivate microalgae

Microalgae strain	Purpose	Wastewater concentration (mg/L)				Pretreatment methods	Metabolic mode	Cultivation mode	Biomass	Nutrients removal	References
		TOC	TN	NH_4^+-N	TP						
Chlorella vulgaris; Chlorella pyrenoidosa; Chroococcus sp.1; *Chroococcus* sp.2	Pollutants removal and biomass production	2965	160.67	201.67	6-9	Remove suspended solid trough centrifugation; diluted	Mixotrophic	Batch/ continue	0.076 g/L/d; 0.12 g/L/d(with glucose)	TP:60%; TN:80%	Prajapati et al., 2014
Chlorella protothecoides	Biofuel feedstock	-	680-1060	123- 385	-	Remove suspended solid trough centrifugation; diluted	Heterotrophic	Bacth/continue	0.70 g/L/d	-	Kobayashi et al., 2013
Scenedesmus quadricauda; Chlorella vulgaris	Treatment of cattle manure	3141-5433		164-273	34-35			Bacth/continue	-	TOC: 38.3%-46.5%; NH_4^+-N: 9.1%-47.2%; TP: 19%-36%	Córdoba et al., 1995 (a); Córdoba et al, 1995 (b)
Neochloris oleoabundans	Nutrient removal and biodiesel feedstock production	-	15-75.2	10.5-52.4	-	No pretreatment	Mixotrophic	Batch	0.03-0.08 g/L/d	NH_4^+-N:90%-95%	Levine et al., 2011
Chlorella vulgaris	Nutrient reduction microalgal biomass production for biofuel and other commodities	COD 516-891	53.7-86.1	50.4-77.7	5.58-13.3	No pretreatment	Mixotrophic	Semi-continuous in lab scale	0.03-0.09 g/L/d	NH_4^+-N: 99.7%-100%; TN: 89.5%-93.6%; TP: 89.2%-92.0%; COD: 55.4%-75.5%	Wang et al., 2010
Chlorella sp.	biofuel	950-2376 COD		90-223.2	10-25	Remove suspended solid trough centrifugation; diluted	Mixotrophic mode	Batch	0.07-0.08 g/L/d	NH_3-N: 100%; TN: 75.7-82.5%; TP: 62.5-74.7%; COD: 27.4-38.4%	Wang et al., 2010
Microalgae and bacteria (no mention of microalgae strains)	Recover nutrients and dry matter/crude protein yields	-	0.64-1.03 g/m²/d	-	-	Filtrated; diluted	Mixotrophic	semi-batch mode	5.3-5.5 g/m²/d	COD: 77%-95%; TN: 39%-62%; TP: 51%-93%	Wilkie et al., 2002
-	Algal productivity and recovery of manure nutrients	300-2500 mg/m²/d			80-420 mg/m²/d	No pretreatment	Mixotrophic	Pilot-scale; Algal turf scrubber raceways	Lowest loading rate (0.3 mg/m²/d):2.5 g/m²/d The highest loading rate (2.5mg/m²/d): 25 mg/m²/d	TN: 51%-83%; TP: 62%-91%	Mulbry et al., 2008

Poultry anaerobic digestion stream used for microalgae cultivation

Table 4. Typical studies about using poultry anaerobic digestion stream to cultivate microalgae

Microalgae strain	purpose	Wastewater concentration (mg/L)				Pretreatment	Metabolic mode	Cultivation mode	Biomass	Nutrients removal	References
		TOC	TN	NH_4^+-N	TP						
Chlorella minutissima; *Chlorellasorokiniana*; *Scenedesmusbijuga*	Animal feed supplementand Biofuel feedstock	-	72-152		6-9	Remove suspended solid trough centrifugation; diluted	Mixotrophic mode	Batch/continue	0.076 g/L/d; 0.12 g/L/d(with glucose)	TP:60%; TN:80%	Singh et al., 2011; Bhatnagar et al., 2011
Chlorella protothecoides	Biofuel feedstock	-	1060; 680	123; 385		Remove suspended solid trough centrifugation; diluted	Heterotrophic mode	Bacth/continue	0.70 g/L/d	–	Liang et al., 2013

It can be seen from these three tables that the purpose of using animal and poultry anaerobic digestion stream to cultivate microalgae are to recovery the nutrients, especially the nitrogen and phosphorus from these wastewater and meanwhile realize biomass production or lipid concentration in microalgae. The common characteristics can be summarized as follows:

(1) Microalgae strain

Microalgae strains are generally sensitive to different types of wastewater due to the imbalance in nutrient profile, deficiency of some important trace elements, and presence of inhibiting/toxic compounds in wastewater streams, and only limited number of strains within a few species could adapt well in different wastewater environments.

It can be seen from the above tables that the tuff microalgae used in animal and poultry manure anaerobic digestion effluent were Chlorophytes (green algae), in this kind of microalgae, *Chlorella sp.* has been used in numerous studies and shown to be effective in removing nitrogen and phosphorus from different wastewater streams with a wide range of initial concentrations. Another chlorophyte widely used for nutrient removal studies is *Scenedesmus sp.*, as mallnon-motile green algae, often clustered in colonies consisting of 2, 4, 8, 16 or 32 cells.

Generally, the microalgae adapted to culture conditions similar to where they were found and generally grew better than those purchased from algae banks (Zhou et al., 2011; Sheehan et al., 1998; Zhou et al., 2012; Zhou et al., 2012; Li et al., 2010; Pérez et al., 2004). Hence, there is a great need to select more robust microalgal strains that are tolerant to specific type of wastewater of interest. Another approach is to select microalgae consortium (a mixed culture of different wild algae species), because it was found that the microalgae consortium performed better than monoculture in terms of nutrient removal and biomass productivity (Chinnasamy et al., 2010; Woertz et al., 2009). Resistant strains can be obtained through genetic engineering and/or breeding manipulation in order to obtain extra resistance to environment stress and/or improve oil synthesis (Uduman et al., 2010; Salim et al., 2010). Genetic and metabolic engineering may have the greatest impacts on improving the economics of production of microalgal biodiesel in the near future (Chisti et al., 2007; Roessler et al., 1994; Dunahay et al., 1996; Zhang et al., 1996).

(2) Influence factors

There are some environment factors that influence the growth of microalgae, such as initial nutrients concentration, light intensity, temperature and inoculation density. Most researches focused on initial nutrients concentration and light intensity. These two factors also directly impact on nutrients removal. In anaerobic digestion stream, the main nutrients are nitrogen and phosphorus, which are also very important nutrients for microalgae strain growth. Hence, the optimal N/P ratio could influence the microalgae strain growth and nutrients removal (Kunikane et al., 1984). However, the optimal N/P and the absolute initial concentration of N or P differ by microalgae species. The rate of phosphorus removal was proportional to the initial nutrient concentration (Rhee et al., 1978; Laliberte et al., 1997). Generally, the uptake rates of ammonium and phosphorus in microalgae were observed to drastically decrease at extremely high concentrations, possibly due to the excessive production of chlorophyll and subsequent light limitation due to self-shadowing (Aslan et al., 2006).

For biomass production, the metabolic mode are very important to the composition of microalgae and then next the nutrients removal. In wastewater treatment processing, microalgae usually carry out heterotrophic or mixotrophic mode to remove the organic carbon, nitrogen and phosphorus. While for inorganic substances, the auto photosynthetic mode also removes the inorganic nitrogen and phosphorus. Besides, the additional carbons sources, such as glycerol or glutamic acid, caused an increase in the uptake of nitrogen (Khan et al., 2008).

(3) Experiment scales

All the above results showed that most of the experiments were carried out in batch and lab-scale. Few of the test and experiments were carried out with semi-continues and large or middle-scale. For the batch and small scale experiment, the initial concentrations of nutrients are relatively lower than that of the continuous and middle scale mode. Most of the continuous and middle scale mode studies were carried out in open ponds with the algal-bacteria or microalgae community system. This indicated that in order to achieve higher nutrients removal, the bacteria and the common effect of many microalgae strains have stronger resistance to higher nutrients concentration and may take main effect in nutrients removal. Hence, the further studies might adopt these systems according to the aim of the studies.

3. Microalgae Cultivation System of Animal and Poultry Digestion Stream

In recent studies about using animal and poultry digestion stream to cultivate microalgae, there are two main cultivation methods: suspension and immobilization. For suspension cultivation mode, the activity of microalgae was very high, and it is easier to control the reaction conditions for microalgae cultivation; for immobilization, the nutrients removal efficiency is higher than that in suspension cultivation systems, but the harvest of microalgae is harder than that in suspension systems.

For suspension systems, there are three mainly systems: The first is the open systems. The raceway pond, atypical open system for algae cultivation, has been extensively used since the 1950s. A raceway pond is usually only about 0.3m deep to provide sufficient sunlight to allow photosynthesis by microalgal cells; the second is the closed systems. Compared with open ponds, the design of closed reactors helps avoid water evaporation and contamination, and increases photosynthesis efficiency. Typical closed reactors include flatplate reactors, tubular photo bioreactors and bag systems (Borowitzka et al., 1999). The third is hybrid systems. Hybrid systems combine the merits of open and closed systems in a two-stage cultivation system. The first stage uses closed photobioreactors to culture the inoculum for the second stage where algae recultivated in open ponds.

In most of the published studies were carried out in small lab scale with flasks. They were conducted with the optimized cultivation conditions such as optimized light intensity and temperature. Only a few studies demonstrated the feasibility of large-scale wastewater based microalgal production in outdoor environment. In order to treat large amounts of wastewater generated from animal and poultry anaerobic digestion stream, bioreactors should be easily scaled up, and operated for efficient nutrient removal. In some other literatures, the cultivation systems reported in large scale include raceway ponds with paddle-wheel agitation, multi-layer open pond-like bioreactors and different types of closed bioreactors such as tubular bioreactor, flat panel bioreactor, coil bioreactor, bag bioreactor, floating bioreactor and solid media surface cultivation bioreactor (Figure 2) (Chisti et al., 2007; Zhou et al.,2012; Oswald et al., 1957;Park et al., 2011; Min et al., 2011; Hu et al., 2013). Wastewater treatment plants may employ high stabilization pond, lagoon and aerated ponds, where algae may also be cultivated. In these bioreactors, multi-layer bioreactor and raceway pond with paddle-wheel were considered as most feasible and cost effective culture systems (Min et al., 2011; Hu et al., 2013). A 2000L and 40,000L pilot-scale multi-layer bioreactors were successfully developed (Min et al., 2011; Min et al., 2013) and were used to cultivate microalgae on animal manure wastewaters for effective algal biomass production and efficient nutrient removal. Wastewater based algal production systems need to be further improved in order to become more competitive and more economically feasible.

Figrue 1. The microalge cultivation systems used (Zhou et al., 2014).

Algae Immobilization

In order to prevent the microalgae moving freely and prevent them wash out from the system, immobilization of algal cells was usually used. Besides, higher nutrients removal efficiency will be obtained in microalgae immobilization system (Hoffmann et al., 1998). There are six types of immobilization methods (covalent coupling, affinity immobilization, adsorption, confinement in liquid-liquid emulsion, capture behind semi-permeable membrane, entrapment) and five types of bioreactors (fluidized-bed, packed-bed, parallel-plate, air-lift, hollow-fiber) (Mallick et al., 2002). Most algal immobilization research was carried out at lab-scale with entrapment method (Mallick et al., 2002). As it was described above, the nutrients removal rate and phosphorus of the immobilized system were higher than that of suspended system. There might be two reasons. The first is the assimilation of microalgae, the second is might to be the absorption of the immobilized materials. However, the biomass growth and production was lower than that in suspended system (Hoffmann et al., 1998). That might be caused by the reduced light which caused by the weaken penetrating of the immobilized materials (Marin et al., 2002). Besides, high cost is also a limiting factor for wide application of immobilization technologies in wastewater treatment.

4. PROSPECTIVE

Although the ability of microalgae to assimilate the excess nutrients from the anaerobic digestion has been studied for many years, there are lots of aspects that need to be further studied.

The first is the microalgae strain selection. In summary, the ideal candidate wastewater-grown microalgae should have following characteristics: (1) fastgrowth; (2) highoil content; (3) high resistance to contamination for different type of wastewater; (4) high tolerance variation of local climate as well as operating conditions. Finally, screening robust wild types trains from local environment and constructing engineered microalgae with desired characteristics may be particularly important and deserve further investigation for advanced wastewater-based algae cultivation system.

The second is the proper pretreatment methods for decreasing the nutrients concentration of anaerobic digested effluent to a proper level for microalgae cultivation. The initial concentration of anaerobic digested effluent was very high which exceed the demand of microalgae and then produced the inhabitation effect to microalgae. Most studies using dilution to relieve the inhabitation effect. However, this method will bring the waste of water resource. Hence, to develop suitable pretreatment methods for the microalgae cultivation by using anaerobic digested stream can promote the biomass production and nutrients recovery.

The third is the cultivation parameters optimization. Although the characteristics of anaerobic digested steams are very different from each other, the nutrients are organic carbon, all kinds of forms of nitrogen and phosphorus which provide microalgae cultivation. Microalgae must carry out the heterotrophic mode or mixotrophic mode and sometimes combing the auto photosynthetic mode to uptake the nutrients in anaerobic digested stream. Hence, optimizing the environmental parameters and combining heterotrophic and mixotrophic cultivation, which is a possible avenue for obtaining maximal algal biomass and lipid productivity, also need to be studied.

Due to the complex characteristics of different kinds of anaerobic digested wastewater, the tests of growing algae in wastewater are mostly at laboratory scale. Pilot-scale algae cultivation continues to face many issues including contamination, inconsistent wastewater components, and unstable biomass production. The major challenge associated with culturing algae in nutrient-rich, natural water bodies comes from the design of the cultivation system. Further research is needed to identify algae species and optimize operating parameters for biomass production, lipid production and nutrients absorption.

The microalgae collection is also another very important issue for microalgae utilization in the following steps that produced from the anaerobic digested stream. The usually trends for the collected microalgae are biodiesel production, biofuel production or fertilizer production. However, the methods used for microalgae collection is costive. How to develop economic and convenient methods for microalgae collection need to be further investigated.

Overall, using microalgae to recover the nutrients from anaerobic digested stream is a prospective technology. It can realize wastewater resource recovery, useful biomass production and further realize bioenergy production. It is a sustainable and may be a replaceable wastewater treatment in the future.

REFERENCES

An, J.Y., Sim, S.J., Lee, J.S., Kim, B.W., (2003). Hydrocarbon production from secondarily treated piggery wastewater by the green alga *Botryococcus braunii*. *J. Appl. Phycol.* 15, 185-191.

Aslan, S., Kapdan, I.K., (2006). Batch kinetics of nitrogen and phosphorus removal from synthetic wastewater by algae. *Ecol. Eng.* 28, 64-70.

Barker, J.C., Zublena, J.P., Walls, F.R., (2001) Livestock and poultry manure characteristics.

Barsanti, L., Gualtieri, P., (2014). *Algae: anatomy, biochemistry, and biotechnology*. CRC Press.

Bhatnagar, A., Chinnasamy, S., Singh, M., Das, K.C., (2011). Renewable biomass production by mixotrophic algae in the presence of various carbon sources and wastewaters. *Appl. Energ.* 88, 3425-3431.

Bjornsson, W.J., Nicol, R.W., Dickinson, K.E., McGinn, P.J., (2013). Anaerobic digestates are useful nutrient sources for microalgae cultivation: functional coupling of energy and biomass production. *J. Appl. Phycol.* 25, 1523-1528.

Borowitzka, M.A., (1999). Commercial production of microalgae: ponds, tanks, tubes and fermenters. *J. Biotechnol.* 70, 313-321.

Bradford, S.A., Segal, E., Zheng, W., Wang, Q., Hutchins, S.R., (2008). Reuse of concentrated animal feeding operation wastewater on agricultural lands. *J. Environ.Qual.* 37, S97–115.

Cheng, H., Tian, G., (2013). Identification of a newly isolated microalga from a local pond and evaluation of its growth and nutrients removal potential in swine breeding effluent. *Desalin. Water Treat.* 51, 2768-2775.

Chinnasamy, S., Bhatnagar, A., Claxton, R., Das, K.C., (2010). Biomass and bioenergy production potential of microalgae consortium in open and closed bioreactors using untreated carpet industry effluent as growth medium. *Bioresour. Technol.* 101, 6751-6760.

Chisti, Y., (2007). Biodiesel from microalgae. *Biotechnol. Adv.* 25, 294-306.

Collos, Y., Berges, J.A., (2004). Nitrogen metabolism in phytoplankton, Encyclopedia of Life Support Systems (EOLSS).

Còrdoba, L.T., Hernàndez, E.S., (1995). Final treatment for cattle manure using immobilized microalgae. II. Influence of the recirculation. *Resour. Conserv. Recy.* 13, 177-182.

Còrdoba, L. T., Hernandez, E. S., Weiland, P., (1995). Final treatment for cattle manure using immobilized microalgae. I. Study of the support media. *Resour. Conserv. Recy.*13, 167-175.

Correll, D.L., (1998). The role of phosphorus in the eutrophication of receiving waters: A review. *J. Environ. Qual.* 27, 261-266.

Dunahay, T.G., Jarvis, E.E., Dais, S.S., Roessler, P.G., (1996, January). Manipulation of microalgal lipid production using genetic engineering. In *Seventeenth Symposium on Biotechnology for Fuels and Chemicals* (pp. 223-231). Humana Press.

Garcia, J., Mujeriego, R., Hernandez-Marine, M., (2000). High rate algal pond operating strategies for urban wastewater nitrogen removal. *J. Appl. Phycol.* 12, 331-339.

De Godos, I., Blanco, S., García-Encina, P.A., Becares, E., Muñoz, R., (2009). Long-term operation of high rate algal ponds for the bioremediation of piggery wastewaters at high loading rates. *Bioresour. Technol.* 100, 4332-4339.

De Godos, I., González, C., Becares, E., García-Encina, P.A., Muñoz, R., (2009). Simultaneous nutrients and carbon removal during pretreated swine slurry degradation in a tubular biofilm photobioreactor. *Appl. Microbiol. Biot.* 82, 187-194.

de Godos, I., Vargas, V.A., Blanco, S., González, M.C.G., Soto, R., García-Encina, P.A., ... & Muñoz, R., (2010). A comparative evaluation of microalgae for the degradation of piggery wastewater under photosynthetic oxygenation. *Bioresour. Technol.* 101, 5150-5158.

Hoffmann, J.P., (1998). Wastewater treatment with suspended and nonsuspended algae. *J. Phycol.* 34, 757-763.

Hu, B., Min, M., Zhou, W., Du, Z., Mohr, M., Chen, P., ... & Ruan, R., (2012). Enhanced mixotrophic growth of microalga *Chlorella sp.* on pretreated swine manure for simultaneous biofuel feedstock production and nutrient removal. Bioresour. Technol. 126, 71-79.

Hu, B., Zhou, W., Min, M., Du, Z., Chen, P., Ma, X., ... & Ruan, R., (2013). Development of an effective acidogenically digested swine manure-based algal system for improved wastewater treatment and biofuel and feed production. *Appl. Energ.* 107, 255-263.

Jimenez-Perez, M.V., Sanchez-Castillo, P., Romera, O., Fernandez-Moreno, D., Pérez-Martınez, C., (2004). Growth and nutrient removal in free and immobilized planktonic green algae isolated from pig manure. *Enzyme Microb. Tech.* 34, 392-398.

Khan, M., Yoshida, N., (2008). Effect of L-glutamic acid on the growth and ammonium removal from ammonium solution and natural wastewater by *Chlorella vulgaris* NTM06. *Bioresour. Technol.* 99, 575-582.

Kobayashi, N., Noel, E.A., Barnes, A., Watson, A., Rosenberg, J.N., Erickson, G., Oyler, G.A., (2013). Characterization of three Chlorella sorokiniana strains in anaerobic digested effluent from cattle manure.*Bioresour. Technol.* 150, 377-386.

Kuenzler, E.J., (1965). Glucose-6-phosphate utilization by marine algae. *J. Phycol.* , 1(4), 156-164.

Kumar, M.S., Miao, Z.H., Wyatt, S.K., (2010). Influence of nutrient loads, feeding frequency and inoculum source on growth of Chlorella vulgaris in digested piggery effluent culture medium. *Bioresour. Technol.* 101, 6012-6018.

Kunikane, S., Kaneko, M., Maehara, R., (1984). Growth and nutrient uptake of green alga, Scenedesmus dimorphus, under a wide range of nitrogen/phosphorus ratio—I. Experimental study. *Water Res.* 18, 1299-1311.

Laliberte, G., Lessard, P., De La Noüe, J., Sylvestre, S., (1997). Effect of phosphorus addition on nutrient removal from wastewater with the cyanobacterium Phormidium bohneri. *Bioresour. Technol.* 59, 227-233.

Larson, W.E., (1979). Crop residues: Energy production or erosion control. *J. Soil Water Conserv.* 34, 74-76.

Levine, R.B., Costanza-Robinson, M.S., Spatafora, G.A., (2011). Neochloris oleoabundans grown on anaerobically digested dairy manure for concomitant nutrient removal and biodiesel feedstock production. *Biomass Bioenergy.* 35, 40-49.

Liang, G., Mo, Y., Zhou, Q., (2013). Optimization of digested chicken manure filtrate supplementation for lipid overproduction in heterotrophic culture *Chlorella prototothecoides*. *Fuel.* 108, 159-165.

Lindstrom, M.J., Gupta, S.C., Onstad, C.A., (1981). Crop residue removal and tillage-effects on soil erosion and nutrient loss in the corn belt.

Xin, L., Hong-ying, H., Jia, Y., (2010). Lipid accumulation and nutrient removal properties of a newly isolated freshwater microalga, *Scenedesmus sp.* LX1, growing in secondary effluent. *New Biotechnol.* 27, 59-63.

Maestrini, S.Y., Robert, J.M., Leftley, J.W., Collos, Y., (1986). Ammonium thresholds for simultaneous uptake of ammonium and nitrate by oyster-pond algae. *J. Exp. Mar.Bio. Eco.* 102, 75-98.

Mallick, N., (2002). Biotechnological potential of immobilized algae for wastewater N, P and metal removal: a review. Biometals, 15, 377-390.

Martinez, M.E., Jimenez, J.M., El Yousfi, F., (1999). Influence of phosphorus concentration and temperature on growth and phosphorus uptake by the microalga Scenedesmus obliquus. *Bioresour. Technol.* 67, 233-240.

Millmier, A., Lorimor, J.C., Hurburgh Jr, C.R., Fulhage, C., Hattey, J., Zhang, H., (2000). Near-infrared Sensing of Manure Ingredients. *Transactions of the ASAE.* 43, 903-908.

Min, M., Hu, B., Mohr, M., Shi, A., Ding, J., Sun, Y., ... & Ruan, R., (2013). Swine manure-based pilot scale algal biomass production system for fuel and wastewater treatment-A case study. *Appl. Biochem. Biotehnol.* 10.

Min, M., Hu, B., Mohr, M.J., Shi, A., Ding, J., Sun, Y., ... & Ruan, R., (2014). Swine Manure-Based Pilot-Scale Algal Biomass Production System for Fuel Production and Wastewater Treatment—a Case Study. *Appl. Biochem. Biotehnol.* 172, 1390-1406.

Min, M., Wang, L., Li, Y., Mohr, M. J., Hu, B., Zhou, W., ... & Ruan, R., (2011). Cultivating *Chlorella sp.* in a pilot-scale photobioreactor using centrate wastewater for microalgae biomass production and wastewater nutrient removal. *Appl. Biochem. Biotehnol.* 165, 123-137.

Molinuevo-Salces, B., García-González, M.C., González-Fernández, C., (2010). Performance comparison of two photobioreactors configurations (open and closed to the atmosphere) treating anaerobically degraded swine slurry.*Bioresour. Technol.*,101, 5144-5149.

Morris, I., Syrett, P.J., (1963). The development of nitrate reductase in *Chlorella* and its repression by ammonium. Arch. Microbiol. 47, 32-41.

Mulbry, W., Kondrad, S., Buyer, J., (2008). Treatment of dairy and swine manure effluents using freshwater algae: fatty acid content and composition of algal biomass at different manure loading rates. *J. Appl. Phycol.* 20, 1079-1085.

Mulbry, W., Kondrad, S., Pizarro, C., Kebede-Westhead, E., (2008). Treatment of dairy manure effluent using freshwater algae: algal productivity and recovery of manure nutrients using pilot-scale algal turf scrubbers.*Bioresour. Technol.* 99, 8137-8142.

Olguín, E.J., Galicia, S., Mercado, G., Pérez, T., (2003). Annual productivity of *Spirulina* (*Arthrospira*) and nutrient removal in a pig wastewater recycling process under tropical conditions. *J. Appl. Phycol.* 15, 249-257.

Oswald, W.J., Gotaas, H.B., Golueke, C.G., Kellen, W.R., Gloyna, E.F., Hermann, E.R., (1957). Algae in waste treatment [with discussion]. *Sewage and Industrial Wastes.* 437-457.

Park, J.B.K., Craggs, R.J., Shilton, A.N., (2011). Wastewater treatment high rate algal ponds for biofuel production. *Bioresour. Technol.* 102, 35-42.

Jimenez-Perez, M.V., Sanchez-Castillo, P., Romera, O., Fernandez-Moreno, D., Pérez-Martınez, C., (2004). Growth and nutrient removal in free and immobilized planktonic green algae isolated from pig manure. *Enzyme Microb. Tech.* 34, 392-398.

Prajapati, S.K., Choudhary, P., Malik, A., Vijay, V.K., (2014). Algae mediated treatment and bioenergy generation process for handling liquid and solid waste from dairy cattle farm. *Bioresour. Technol.*,167, 260-268.

Ragauskas, A.J., Williams, C.K., Davison, B.H., Britovsek, G., Cairney, J., Eckert, C.A., ... & Tschaplinski, T., (2006).The path forward for biofuels and biomaterials. Science, 311, 484-489.

Rhee, G.Y., (1978). Effects of N: P atomic ratios and nitrate limitation on algal growth, cell composition, and nitrate uptake 1. *Limnol. Oceanogr.* 23, 10-25.

Roessler, P.G., Brown, L.M., Dunahay, T.G., Heacox, D.A., Jarvis, E.E., Schneider, J.C., ... & Zeiler, K.G., (1994, January). Genetic engineering approaches for enhanced production of biodiesel fuel from microalgae. InACS symposium series (Vol. 566, pp. 255-270). Washington, DC: American Chemical Society,[1974]-.

Ruiz-Marin, A., Mendoza-Espinosa, L.G., Stephenson, T., (2010). Growth and nutrient removal in free and immobilized green algae in batch and semi-continuous cultures treating real wastewater. *Bioresour. Technol.* 101, 58-64.

Salim, S., Bosma, R., Vermuë, M. H., Wijffels, R.H., (2011). Harvesting of microalgae by bio-flocculation. *J. Appl. Phycol.* 23, 849-855.

Dunahay, T., Benemann, J., Roessler, P., (1998). A look back at the US Department of Energy's aquatic species program: biodiesel from algae (Vol. 328). Golden: National Renewable Energy Laboratory.

Singh, M., Reynolds, D.L., Das, K.C., (2011). Microalgal system for treatment of effluent from poultry litter anaerobic digestion. *Bioresour. Technol.* 102, 10841-10848.

Travieso, L., Sánchez, E., Borja, R., Benítez, F., León, F., Colmenarejo, M.F., (2004). Evaluation of a laboratory and full-scale microlage pond for tertiary treatment of piggery wastes. *Environ. technol.* 25, 565-576.

Uduman, N., Qi, Y., Danquah, M.K., Forde, G.M., Hoadley, A., (2010). Dewatering of microalgal cultures: a major bottleneck to algae-based fuels. J. Renew. Sustain. Ener. 2, 012701.

Wang, L., Li, Y., Chen, P., Min, M., Chen, Y., Zhu, J., Ruan, R.R., (2010). Anaerobic digested dairy manure as a nutrient supplement for cultivation of oil-rich green microalgae *Chlorella sp. Bioresour. Technol.* 101, 2623-2628.

Wang, L., Wang, Y., Chen, P., Ruan, R., (2010). Semi-continuous cultivation of *Chlorella vulgaris* for treating undigested and digested dairy manures. Appl. Biochem. Biotech. 162, 2324-2332.

Wilkie, A.C., Mulbry, W.W., (2002). Recovery of dairy manure nutrients by benthic freshwater algae. *Bioresour. Technol.* 84, 81-91.

Woertz, I., Feffer, A., Lundquist, T., Nelson, Y., (2009). Algae grown on dairy and municipal wastewater for simultaneous nutrient removal and lipid production for biofuel feedstock. J. Environ. Eng. 135, 1115-1122.

Yetilmezsoy, K., Sakar, S., (2008). Development of empirical models for performance evaluation of UASB reactors treating poultry manure wastewater under different operational conditions. *J. Hazard. Mater.* 153, 532-543.

Zhang, Z., Moo-Young, M., Chisti, Y., (1996). Plasmid stability in recombinant Saccharomyces cerevisiae. *Biotechnol. Adv.* 14, 401-435.

Zhou, W., Li, Y., Min, M., Hu, B., Chen, P., Ruan, R., (2011). Local bioprospecting for high-lipid producing microalgal strains to be grown on concentrated municipal wastewater for biofuel production. *Bioresour. Technol.* 102, 6909-6919.

Zhou, W., Chen, P., Min, M., Ma, X., Wang, J., Griffith, R., ... & Ruan, R., (2014). Environment-enhancing algal biofuel production using wastewaters. *Renew. Sust. Energ. Rev.* 36, 256-269.

Zhou, W., Li, Y., Min, M., Hu, B., Zhang, H., Ma, X., ... & Ruan, R. (2012). Growing wastewater-born microalga Auxenochlorella prototothecoides UMN280 on concentrated municipal wastewater for simultaneous nutrient removal and energy feedstock production. *Appl. Energ.* 98, 433-440.

Zhou, W., Min, M., Li, Y., Hu, B., Ma, X., Cheng, Y., ... & Ruan, R. (2012). A hetero-photoautotrophic two-stage cultivation process to improve wastewater nutrient removal and enhance algal lipid accumulation. *Bioresour. Technol.* 110, 448-455.

PART III. SCALE-UP AND INDUSTRIALIZATION

In: Gas Biofuels from Waste Biomass
Editor: Zhidan Liu

ISBN: 978-1-63483-192-5
© 2015 Nova Science Publishers, Inc.

Chapter 10

INDUSTRIAL CASE STUDY: BOTTLENECKS AND PERSPECTIVES

Li Jian[*], *Luo Guanghui and Sheng Liwei*
Heilongjiang Institute of Agriculture Mechanical Engineering Science,
Harbin, China

ABSTRACT

The first garage type dry fermentation project in China was constructed in May 2013. For expounding the first Chinese industrialization practice of dry fermentation technology, this chapter introduces the project's core technology and implements the situation in detail.

As the first Chinese industrialization practice of dry fermentation technology, Bin County garbage disposal comprehensive utilization project could produce bio-methane for vehicles. There are great expanding spaces from the raw material collection and the products utilized. Relatively, the bottlenecks of the garage type dry fermentation project still exist, including difficult homogeneity, low stability, biogas waste etc. but considering the long-term operation and environment protection, promotion of dry fermentation technology is necessary.

Keywords: MSW, treatment, garage, dry fermentation

1. INTRODUCTION

Anaerobic technology is a great invitation to produce new energy, which offers us a way to resolve the shortage of traditional fossil energy. As we know, the anaerobic process uses a different kind of bacteria to produce CH_4 and CO_2 in suitable water, temperature and anaerobic environments. The target of anaerobic technology is to produce CH_4 (the main component of natural gas) for optimizing the anaerobic process. To achieve this, we should

[*] Corresponding author: 323#, No. 156, Haping road, Nangang district, Harbin city, Heilongjiang province, PRC. Email: lijian_0498@163.com.

control different factors, including the digestion of material, bacteria, O_2, temperature, pH, TS, mix, etc. We can classify the anaerobic technology to wet fermentation and dry fermentation according to TS concentration.

There are many processes for wet and dry fermentation technology; wet fermentation processes include CSTR, HCPF, USR, UASB, AF, UBF etc. while dry fermentation processes include Garage, Dranco, Kmpogas, Valorga etc. Dry fermentation technology was started in Europe during the 1980s; the basic theory uses digested water to mix bacteria and fresh material instead of traditional mechanical mix. The dry fermentation could save more clean water and reduce more waste water. The most economic digestion temperature of a biogas project is 35°C -38°C, which needs large amounts of heat to keep the moderate temperature. Saving more water means saving more energy, which is the real reason why we pay more and more attention to dry fermentation technology.

On July 7th 2010, Heilongjiang Longneng Weiye Gas Shares CO.,LTD., German Technology Cooperation Company(GIZ), Heilongjiang Institute of Agriculture Mechanical Engineering Science and Heilongjiang Province Rural Energy Office signed the contract of SINO-German 'international best practice' demonstration biogas project ("PPP-Development of SINO-German 'international best practice' Demonstration biogas plants under the aspect of innovations"), which requested GIZ to supply the technical service to Heilongjiang Longneng Weiye Gas Shares CO.,LTD. for free. GIZ then authorized German K&F biogas design company to design the detail anaerobic process, Heilongjiang Institute of Agriculture Mechanical Engineering Science should use Chinese Agriculture Department 948 Project to import automatic control system and biogas purification technology and Heilongjiang Province Rural Energy Office support the relevant policies. The demonstration project started around May 2011, which is the first garage type dry fermentation project in China. Now, we will show you more detail about this project.

2. PROJECT INFORMATION

2.1. Basic Information

Name of project: Bin County MSW Integrated Treatment Project
Address of project: located in Xinli village, Binzhou town, Bin county, Harbin city, Heilongjiang province, China. The nearest resident is 1500m away from there, total floor area is 33025 m^2.
Operation company of project: Heilongjiang Longneng Weiye Gas Shares CO.,LTD.
Scale of project: MSW 58,400t/y, bio-methane 9278m³/d.
Investment of project: RMB 43,259,800 yuan.

2.2. Core Process of Project

The process of project could be divided into 4 parts: pretreatment part, anaerobic part, fertilization part and purification part.

The first part: MSW pretreatment.

All MSW from Bin County is delivered to the storage area of a pretreatment hall by government garbage truck. We then use a wheel loader to put the MSW into a slat type spreader. When the MSW is well-distributed on the transfer belt, four workers pick the stone, glass bottle, and big metal while the other MSW is delivered into the shredder. Then the MSW is crushed smaller than 80mm and delivered into a rotary drum and the waste inside of rotary drum is crushed again and made to be RDF. The waste outside of rotary drum is delivered to receive a box before blowing the small plastic pieces out.

The second part: MSW anaerobic treatment.

We deliver the bio waste into the garage type dry fermenter by wheel loader. There are twelve fermenters, and the dimensions of each fermenter is 24m×4m×4.2m. The height of the waste in the fermenter is about 3 meters; that means we can load 276m³ waste per fermenter. After fulfilling this, we fix the baffle and close the hydraulic door, then start the spray system. The oxygen in the fermenter is exhausted by aerobic bacteria in 1 to 2 days. After that, the bio waste turns into the anaerobic process. When the methane concentration is up to 30%, we start collecting the biogas. At the last time of anaerobic process, you can find the flow of biogas is under 5m³/h. After this, we start the waste gas exchange system; when the methane concentration is lower than 4%, we use an air blower to blow the fresh air into fermenter. This time, the wheel loader appears again the same way to take the digested waste out. As the wheel loader works, the fresh air blower keeps working too.

The third part: digested MSW treatment.

The digested waste in the fermenter is delivered to the fertilizer hall by a wheel loader. We use the self-propelled turning machine to dry the waste until the water content is under 70%. Once that is done, we can use the fine rotary drum to sieve the dry waste into fertilizer and other waste, which will be sent to a landfill.

The fourth part: Biogas purification.

Biogas from the fermenter is pumped into a gas holder while the the pressure of gas increase to 30 mbar and use desulfurizer to absorb the H_2S. After that, the pressure increase of gas to 6 bars by the biogas compressor and is then put into ammonium hydroxide absorb tower to absorb CO_2. Lastly, the bio-methane will be sent into nature gas station.

3. Project Implementing Situation

3.1. MSW Treatment Hall

The MSW treatment hall is the most important part of the whole project. There are twelve fermenters with dimensions of 24m×4m×4m, the volume is 384m3; there is a percolation buffer pool, with dimensions of 6.85m×2.5m×3.5m. Two percolation tanks are used with the dimension is 24m×4m×5.5m and a volume of 528m³. The process of leaking waste water is flowed from the button of the fermenter to the percolation buffer pool. We then use a suck pump to suck the waste water into the percolation tank. At last, we use the spray pump to suck the waste water in the percolation tank into each fermenter.

(a)

(b)

(c)

Figure 1. Construction situation of MSW treatment hall. (a) The outside of MSW treatment hall. (b) The inside of MSW treatment hall. (c) The top of the dry fermenters.

3.2. Process Control Room

The process control room is the core which monitors and controls each part of the whole project, especially the digestion process. We imported a suit of control systems from Biogas Process Control Company of Sweden. The control point includes temperature of the mass, flow, pressure, component of the biogas, temperature, liquid level of digested water, temperature of heating pipe, pressure of seal bag of door, offline pH test, and methane leakage alarm. These parameters will be the theory procedure of controlling the electric valve, pump, and blower to reach the best conditions.

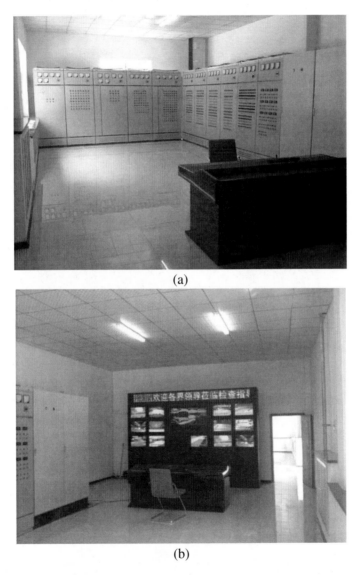

(a)

(b)

Figure 2. Construction situation of process control office. (a) The inside of control room. (b) The inside monitoring system of control room.

3.3. The Other Auxiliary Facilities

The other auxiliary facilities include a boiler, biogas holder, purification hall and flare. There is a DZL/S1.4-0.7/95/70 pellet boiler, which can supply the heat to the whole system. The volume of the gas holder is 600m³ and the pressure is under 1KPa. The function of the gas holder is to buffer the purification system. The flare's max burning flow is 500m³/h, which could burn all biogas should an emergency situation occur.

(a)

(b)

(c)

Figure 3. (Continued).

(d)

Figure 3. Construction situation of other auxiliary facilities. (a) The boiler room. (b) The gas holder. (c) The biogas purification room. (d) The flare.

Table 1. Index of the energy conservation and emission reduction

Index	value		note
Bio-methane	3,386,470	m³/y	
Petrol replacement	2466.6	t	
Coal conservation	3628.2	t	
CO reduction	25.31887	t	97%
NOx reduction	21.67712	t	39%
HC reduction	19.98783	t	72%
Lead compound reduction	0.30603	t	100%
CO2 reduction	9043.278	t	24%
SO2 reduction	272.1147	t	90%
Dust reduction	2467.173	t	
Plastic and paper recycle	4978.6	t	
Metal recycle	146	t	
Glass recycle	693.5	t	

4. TARGET OF ENERGY CONSERVATION AND EMISSION REDUCTION

This project could produce bio-methane 9278m³/d, according to calculations, each index of the energy conservation and emission reduction is showed in table1.

We can get some important parameters from above table. This project can produce 3,386,470 cubic meter bio-methane per year, which is equal to the supply of 3,970,000 liters of 93# petrol. This means we can reduce CO_2 and SO_2 9043.278 t and 272.1147 t per year respectively. Based on this data, we believe the dry fermentation technology will be the best way to dispose municipal solid waste.

CONCLUSION AND SUGGESTION

1) The wet and dry fermentation technologies are different ways to dispose organic waste. Although the dry fermentation project needs a large initial investment and high standard automatic control system, the long-term operation and promotion of environment protection makes dry fermentation technology more favorable.

2) For the large—even the super large—solid waste treatment project, the concept design should stay the same with the local government. The best cooperation mode should be like this: the government is responsible for investment and construction, while the private company is responsible for management and operation. The private company has to keep the best operation state to obtain maximum profit. Only that cooperation mode could effectively avoid the risk of these kinds of projects.

3) As the first Chinese industrialization practice of dry fermentation technology, Bin County garbage disposal comprehensive utilization project could produce bio-methane for vehicles. There are great expanding spaces from raw material collection and the product utilization.

We believe the garage type dry fermentation technology will receive further development in China, especially through the two year stable operation of Bin County garbage disposal comprehensive utilization project.

REFERENCES

[1]   Zhou, Xiujie. Analysis and prospect of biomass energy utilization China investment industry forecast report in 2010-2015[R]. 2010, 1-3

[2]   Liang, Xu; Liang, jinguang. The development and application prospect of China "biogas industry" [J]. *Agricultural Engineering.*, 2010, 26(5), 1-6

[3]   The Ministry of Agriculture Department of personnel and labor. The biogas production (M).Chinese Agriculture Press. 69-77y

[4]   Li, Xiujin. Solid waste engineering [M]. Beijing: Chinese Environment Science Press, 2003

[5]   Chen, Hongzhang; Xu, Jian. Modern solid state fermentation principle and application [M]. Beijing: Chemical Industry Press, 2002

[6]   Wang, Shuying; Gao, Chunti. Environmental introduction [M]. Beijing: Chinese Architecture Industry Press, 2004

[7]   Qu, Jingxia, Jiangyang. Agricultural waste, dry anaerobic fermentation technology [J]. *Renewable Energy.* 2004, 16(2), 40-41

[8]   Kelleher, M. Anaerobic Digestion Outlook for MSW streams [J]. *Biocycle*, 2007, 48(8), 51-55

[9]   Nichols, CE. Overview of anaerobic digestion technologies in Europe [J]. *Biocycle*, 2004, 45(1), 47-54

[10] Laelos, HF; Desbois, S; Saint Joly, C. Anaerobic disgestion of municipal solid organic waste:Valorgafull-scale plant in Tilburg,the Netherlands[J]. *Water Science and Technology*, 1997, 36(6), 457-462

[11] Luning, L; Zundert, EH, Brinkmann, AJ. Comparison of dry and wet digestion for solid waste [J]. *Water Science and Technology*, 2003, 48(4),15-20

[12] Li, Dong; Sun, Yongming. The application research on the technology of anaerobic digestion treatment of city life garbage [J]. *Advances in biomass chemical treatment.*, 2008, 42(4), 43-50

[13] Wu, Manchang; Sun, Kewei. Different reaction temperature China city waste anaerobic fermentation of [J]. *Chemical and biological engineering.*, 2005, 9, 28-31

[14] Lu, Weiya. Application of city environment and city in the treatment of organic garbage ecological. *Anaerobic fermentation technology*, 2002, l5(6), 55-5 7

[15] Li, Chao; Lu, Xiangyang. City organic garbage garage type dry fermentation technology [J]. *Renewable energy.*, 2011, 29 (2) 35-38

EDITOR CONTACT INFORMATION

Dr. Zhidan Liu
Associate Professor
China Agricultural University
17 Tsinguha East Rd, Haidain District, Beijing, 100093, China
zaliu@cau.edu.cn

INDEX

C

D

E

H

I

R